建筑立场系列丛书 No.51

C3

景观与建筑
Landscaping and Building

汉英对照
（韩语版第367期）

韩国C3出版公社 | 编

时真妹 马莉 栾一斐 刘文静 孙雯雯 周一 蒋丽 王单单 | 译

大连理工大学出版社

4 景观与建筑

- 004 建筑、区域与自然：景观一体化还是构建景观？
  _João Pedro T. A. Costa
- 010 安蒂诺里酒庄 _ Archea Associati
- 028 Vučedol考古博物馆 _ Radionica Arhitekture
- 034 克尔科诺谢山环境教育中心 _ Petr Hájek Architekti
- 046 毛利儿童保育及社区中心 _ Collingridge and Smith Architects
- 058 莫斯格史前博物馆 _ Henning Larsen Architects
- 068 葡萄牙奥比多斯科技园中央大楼 _ Jorge Mealha
- 084 小松科技园 _ UAO

96 住宅 立足于气候

- 96 居所与气候 _ Aldo Vanini
- 102 沙漠中的黑色住宅 _ Oller & Pejic Architecture
- 110 沙漠庭院住宅 _ Wendell Burnette Architects
- 120 哈达威住宅 _ Patkau Architects
- 132 四季住宅 _ Casos de Casas

140 自然与人工——一分为二还是合而为一？

- 140 自然与人工——一分为二还是合而为一？ _Paula Melâneo
- 146 Pacific Dazzle美发沙龙 _Atelier KUU
- 154 欢乐餐厅 _H&P Architects
- 162 紫禁城红墙茶室 _CutscapeArchitecture
- 172 Steirereck餐厅 _PPAG Architects
- 180 Angolino餐厅 _Geneto

- 186 建筑师索引

4 Landscaping and Building

004 *Architecture, Territory and Nature: Landscape Integrating or Building Landscapes?*
_João Pedro T. A. Costa

010 Antinori Winery _ Archea Associati

028 Vučedol Archaeological Museum _ Radionica Arhitekture

034 Krkonoše Mountains Environment Education Center _ Petr Hájek Architekti

046 Maori Childcare and Community Center _ Collingridge and Smith Architects

058 Moesgaard Museum _ Henning Larsen Architects

068 Óbidos Technological Park Central Building _ Jorge Mealha

084 Science Hills Komatsu _ UAO

Dwell How
96 Stand on the Climate

96 *Dwellings and Climate _ Aldo Vanini*

102 Black Desert House _ Oller & Pejic Architecture

110 Desert Courtyard House _ Wendell Burnette Architects

120 Hadaway House _ Patkau Architects

132 Seasonless House _ Casos de Casas

140 Natural and Artificial - Dichotomy or Duality?

140 *Natural and Artificial - Dichotomy or Duality?_Paula Melâneo*

146 Pacific Dazzle Baton_Atelier KUU

154 Cheering Restaurant_H & P Architects

162 The Forbidden City Red-wall Teahouse_CutscapeArchitecture

172 Restaurant Steirereck_PPAG Architects

180 Angolino Restaurant_Geneto

186 Index

景观与建筑
Landscaping

  当建筑项目需要融入大型景观空间时，会呈现一种新的维度。无论在农村还是城市，建筑与公共空间不仅 (1) 借助自然大力加强同周围环境之间的关系，同时还要发挥一定的作用；(2) 引入更为宽阔的视野与视角；(3) 考虑人工建筑与土地的敏感关系；(4) 自身与道路、输水管道等长距离建造物之间的位置关系；(5) 明确对实地的影响，权衡运用掩饰的设计手法，促进建筑与景观的相互融合，或者以标新立异的设计姿态，打造全新的景观。

  基于上述五条原则，此类建筑项目大都独具特色。与景观相关联的建筑设计作品也因此面临着特殊的挑战。这些挑战将在本章收录的七个项目中得到详细的探讨。

Architecture projects gain a new dimension when they have to configure large landscapes. Both on rural or on urban areas, buildings and public spaces do not only (1) strongly reinforce its relation with the surroundings through nature; they also have to work; (2) the introduction of larger horizon perspectives; (3) the sensitive relation between the artificial construction and the land; (4) its situation among long structures of pathways, waterlines, etc., and; (5) the definition of site impacts, balancing between dissimulation approaches, promoting its integration on the landscape, or intentional design gestures, creating themselves new landscapes.

These five topics justify why this type of projects present own characteristics. The design of architecture pieces in relation with the landscape faces, therefore, specific challenges, which are illustrated and discussed along the seven following projects.

安蒂诺里酒庄_Antinori Winery/Archea Associati
Vučedol考古博物馆_Vučedol Archaeological Museum/Radionica Arhitekture
克尔科诺谢山环境教育中心_Krkonoše Mountains Environment Education Center/Petr Hájek Architekti
毛利儿童保育及社区中心_Maori Childcare and Community Center/Collingridge and Smith Architects
莫斯格史前博物馆_Moesgard Museum/Henning Larsen Architects
奥比多斯科技园中央大楼_Óbidos Technological Park Central Building/Jorge Mealha
小松科学博物馆_Science Hills Komatsu/UAO

建筑、区域与自然：景观一体化还是构建景观？_Architecture, Territory and Nature: Landscape Integrating or Building Landscapes?/João Pedro T. A. Costa

建筑、区域与自然：景观一体化还是构建景观？

区域内的建筑是公共领域的一部分，因此不能作为单独建筑物来设计，而是作为整个场地中的一部分。同公共空间一样，此类建筑也会成为广阔自然环境中的一景。它们不仅在空间上具有整体性，在时间维度上也是如此。景观建成之后，人类的介入引入了时间维度，景观也因此拥有了自己的生命。它们适应冬夏的交替、温度的变化、颜色的改变、用途的转换，这些都在不断检验着建筑项目的耐久性。时光流逝终将使每个设计方案中冗余奢华的败笔暴露无遗，而精华要素的价值得到凸显。

区域建筑设计决策的每一步都要有整体意识。

一个项目应当如何自然而然地融入周围环境？在大片的广阔区域视野中，项目发挥的作用是什么？人工建筑物与地貌、植被之间关系如何？它如何与行人步道、水流景致融为一体？应当采用何种建筑理念与手段彰显场地特色？

上述第五个问题将下文的七个建筑项目划分为两大类，我们可以简单称之为"景观一体化"类型与"构建景观"类型。

景观一体化意味着这些项目在满足建筑功能需求的同时，力图成为现有生态区域的一部分。它们尽量采用某种方式伪装自己，降低自身的存在对场地造成的影响，但是建筑不可能对场地完全不造成任何影响，因为从概念上来讲，设计方案也并不打算完全隐藏建筑。因此，核心问题是如何实现景观一体化与建筑表现之间的平衡。显然，在理论上这与生态规划中的"设计结合自然"概念密切相关。1969年，苏格兰景观设计师Ian McHarg率先提出"设计结合自然"的概念，这一理论不仅体现在设计方法的选用、土地调控意识方面，还体现在植被的协调一致性方面。

第二类是构建景观，这类项目在塑造场地全新形象的过程中发挥着积极的作用。建筑项目为场地施工中新姿态的确立提供了契机，它既期望被人们主动视为新景观的一部分，同时又期望借助自然同现有景观

Architecture, Territory and Nature: Landscape Integrating or Building Landscapes?

Architecture on the territory is part of public realm. Its design can't be conceived as an isolated object, but as part of a larger site. As well as the public spaces, these buildings become part of a broader natural environment. And if they are collective in space, they are also collective in time. Allowing men's located intervention, landscapes have their own life after built, when the time dimension is introduced. They accommodate successive winters and summers, different temperatures, different colors, different uses, testing the consistence of the project. Implacable, the time dimension clearly denounces the superfluous and valorizes the essential of each design solution.

The architectural design on the territory faces its collective sense in each decision.

How does each project conceive its relation with the surroundings through nature? What role does it assume on the large territorial horizon perspectives? What's the relation between the artificial building construction and the land and vegetation? How does it integrate pedestrian pathways and water flows? What's the conceptual approach to the site characteristics?

This fifth question is used to divide the seven projects in two groups, which can be shortly named as the "landscape integration" group and the "building landscapes" group.

Landscape integration means these projects try to be part of the existing organic territory, despite the need to fulfill the building's program. Somehow, they try to dissimulate themselves and reduce its site impact, although it isn't null – neither the conceptual design approach is to totally hide the building. The balance between landscape integration and located architecture expression becomes, therefore, a central question. The theoretical relation with the "design with nature" concept of ecological planning is evident, as the Scottish landscape architect Ian McHarg pioneered launched it, in 1969, finding its expression on the design approach, on the land modulation sense, and on the vegetation integration. The second group, building landscapes, means these projects assume an active role in the creation of a new site image. The

安蒂诺里酒庄,意大利巴尔吉诺
Antinori Winery in Bargino, Italy

照片提供:©Archea Associati (Pietro Savorelli)

之间保留某种关联。对此类设计案例而言,新建筑物所呈现的积极姿态与它试图同场地所建立的关系——即新老建筑的和谐一致性、如何平衡两者之间的关系便成为一个关键问题。于是形成了一种对话状态,时而作为积极的建筑形象,时而借助自然的掩饰,仿佛项目只想保留"足够"积极的建筑设计来肯定自己,而掩饰了其他建筑元素——因此,建筑的积极形象得到了强化,更易于成为新的地标性建筑。

为说明这种概念性方法,我们将介绍七个分别涉及建筑、区域、自然的项目。其中,前三个是景观一体化设计理念的例证,后四个则展现了构建景观型的设计理念,每个项目都将建筑意图融于设计构思之中,并提供多种设计方案以供讨论。

由Archea Associati设计的安蒂诺里酒庄,位于意大利巴尔吉诺的基安蒂地区,该地区特有的标志性景观是一望无际的葡萄园。这栋建筑对周围景观影响极低,因为其整体掩映在一处山坡中,屋顶完全被农田覆盖,沿着自然延伸的山坡种满成排的葡萄藤。建筑正立面散布的水平切口以及屋顶和地面的若干个圆形开孔,使光线能够深入建筑内部,为室内提供采光。该项目包括:位于最底层的葡萄酒生产及储藏区,位于建筑上层的博物馆、图书馆、礼堂、品酒区和销售区。

这种景观一体化设计方法揭示出建筑与自然之间强烈的联系。项目规划只允许产生较小的外部视觉冲击,因此建筑需要遁形于周围环境之中,仿佛融为山丘的一部分。设计师对建筑材料的选择独具匠心,主体采用陶土材料,搭配混凝土为表层着色,还有耐蒸汽腐蚀的考顿钢及木材。本案作为一种强有力的建筑表现形式——尤其是就其内部立体空间的可塑性而言——与景观一体化设计手段完美结合,使人工作品与自然环境和谐相融。建筑利用土壤作为天然隔离层,以保持稳定的室内气候环境。

位于克罗地亚武科瓦尔的Vučedol考古博物馆,由Radionica Arhitekture建筑事务所设计。建筑能够通向高出地平面20余米的Vučedol

architecture project is the opportunity to affirm a new gesture in the construction of the territory; it wants to be actively seen as part of the new landscape, although simultaneously maintaining some relation with the existing one, by using nature. The balance between the affirmative gesture of the new architecture piece and the type of relation it tries to establish with the site, namely its coherence, becomes the central questions for these cases. A dialog is generated between the moments of affirmative architecture and the moments of natural dissimulation, as if the project just wanted the "enough" affirmative building design, partially dissimulating the other elements of the building – therefore contributing to reinforce the image of the affirmative ones, which tend to constitute new land marks.

To illustrate this conceptual approach, seven different projects of architecture, territory and nature are presented: the first three exemplifying a landscape integration intention and the last four exemplifying a building landscape's one. Carrying intentionality in its conception, each illustrated case allows the discussion of the design options.

The Antinori Winery, in Bargino, Italy, designed by Archea Associati, is located on the special landscape area of Chianti, with its vineyards marking the territory. It is conceived as a low impact building whose body merges with the hillside, with its roof being entirely covered with farmland, cultivated along the natural slope with interrupt vines. Interior lightning is provided by the discrete frontal horizontal cuts and by some circular openings located on the roof and floors, bringing light into its depths. Its program includes, at the lowest level, the wine production and storage facilities and, at the upper level, a museum, a library, an auditorium, and areas for wine tasting and shopping.

This landscape integration approach reveals a very strong relation with nature. The program allows the project to have an exterior small visual impact, dissimulating its presence, as if the building was part of the hill. A special attention was given to the construction materials, dominating terracotta with the combination of concrete pigmentation, corten steam and wood. A strong architecture expression, especially on the interior's plasticity, is very well combined with the landscape integration approach, merg-

莫斯格史前博物馆,丹麦奥尔胡斯
Moesgard Museum in Aarhus, Denmark

文化考古发现遗址。博物馆的设计兼具外部入口坡道和室内考古发现成果坡道展厅的双重功能,因此采用依傍山坡建造的之字形人工作品形态。这样一来,外部空间就掩映在天然形成的屋顶之下,而室内展区呈现出长廊的效果。

在本案中,行人步道成为催生该项目设计的基础元素。虽然屋顶突出的棱角与坡道舒展的曲线产生一定的冲突效果,但是景观一体化主要还是通过坡道上绿植覆盖的屋顶来实现的。由于坡道的正面采用红色陶制材料,因此外墙材料的选择使该建筑设计处于融入场地环境和展现积极姿态的中间状态。混凝土表面或是栏杆上尚未完工的图像标志着建筑的表现形式。该建筑通过半地下建造结构,展现了融入场地的建筑物如何解决项目本身所带来的室内设计与自然采光控制方面的不利状况。

Petr Hájek Architekti建筑师事务所设计的位于捷克共和国弗尔赫拉比市的克尔科诺谢山环境教育中心,坐落在一个城市公园内,后面是国家公园行政大楼。项目规划包括一个展览区、一个图书馆、一个报告厅、一个实验室和一个礼堂,另配有一个停车场。这座半地下建筑的屋顶棱角分明,被绿植覆盖。受山体几何形式的启发,还采用了金属元素。屋顶的构造是对山体形态的小规模再现,不仅如此,教育中心建造的人工边坡也与大自然形成了一种敏感关系。人工边坡位于公园和16世纪古城堡的对面,边坡上的植被处理与公园的树木、绿地在植物适应性、色彩方面均保持一致。建筑出入口的位置都经过精心设置,以便控制室内采光,较大的入口位于面向国家公园行政大楼一侧。建筑外观以自然旧的表现形式干净利落地展现了当代设计风格;在室内设计中,裸露的混凝土屋顶与明快、精致的内部隔断及家具形成了鲜明的对比,前者真实地反映出建筑的构架,而后者则主要采用胶合板材料。土壤发挥天然保温层的作用,因此起到了很好的节能效果。

在景观类设计实例当中,我们主要关注位于日本石川的小松科学博物馆,该馆由UAO设计,坐落在一座工厂旧址上,地处开阔的人造都市

ing the work of man with the natural environment. The building also uses the earth as a natural insulator, to maintain a constant indoor climate.

The Vučedol Archaeological Museum, in Vukovar, Croatia, designed by Radionica Arhitekture, connects the access road to the site of the Vučedol culture archaeological findings, located 20 meters above on a plateau. It combines the double function of exterior ramp access and interior ramp exhibition of the findings, therefore introducing this zigzag artificial form on the hill slope. As a result, exterior spaces are dissimulated by the natural roof and interior exhibition areas present a long corridor form.

In this case, the pedestrian pathway is the generator element of the project. Landscape integration occurs mainly through the green roof of the ramp, although its angular form conflicts with the curvilinear slop of the hill. The exterior material selection, with the presence of the red ceramic ramp front, is halfway between the site integration and an affirmative gesture. Some unfinished image marks the building expression, namely on the concrete's surface or on the handrails. The building shows how site integration objectives, through half buried constructions, may generate difficult situations to solve in the interior design and in the natural light control.

The Krkonoše Mountains Center for Environmental Education, in Vrchlabi, Czech Republic, designed by Petr Hájek Architekti, is located in a city park, in front of the administration building of the National Park. Its program is fulfilled through an exhibition area, a library, a lecture room, a laboratory and an auditory, complemented by a car parking. The half-buried construction is covered by an angular green roof, with the metal vectors inspired in the mountain's geometry. The Center, more than the roof small scale reproduction of the mountain's morphology, finds a sensitive relation with nature through the proposal of an artificial slope facing the park and the 16th century City Castle, treated with the same vegetation, therefore reproducing the plasticity and colors of the park's trees and grass. Its openings are carefully located in order to control the interior lighting, the large one being on the side of the National Park administration building. While the building exterior expression decrepitly assumes a clean contemporary language,

环境之中。博物馆的规划设计包括一个3D影院、一个活动厅、多功能厅、一个展览区、一个科学实验室、研讨会议区、休息室和管理区。建筑整体形象是带弧度的穹顶,上面覆盖绿草,形似起伏的人造山丘,游客可以自由攀爬穿梭其中。

在本案中,建筑通过长弧形线条来掩饰自身的存在,尽管已经尽可能地克制,但是这本身仍是一种积极的建筑姿态。屋顶铺设的绿草弥补了周边环境中绿地的缺失,同时还设有一条公园步道。作为当代建筑,本案的主要表现形式采用裸露的混凝土、玻璃材料和白色喷漆的钢柱。出乎意料的是,项目展现的积极姿态仅限于建筑主体本身,大型防水地上停车场不见一棵树木,并未延续这种积极的姿态。

Henning Larsen Architects建筑师事务所设计的丹麦莫斯格史前博物馆,位于奥尔胡斯市郊区Skåde的丘陵景观区,周围绿地环绕。倾斜的矩形屋顶是该建筑的显著特点,上面铺设草坪,宛如平地长出的一个平台,夏季可作公园供人们休闲使用,冬季降雪时节则成为雪橇滑道。博物馆的室内呈阶梯状台地设计,这一设计灵感来源于考古发掘的过程以及层次分明的历史文明。

虽然部分结构掩映于地下,但是该建筑着意成为区域内全新的视觉地标,建筑占地面积之广、钢筋混凝土结构产生的视觉冲击之大,都成就了这一地标性建筑。倾斜的屋顶景观形成了一个"天然的"观景台,在台上可以一览奥尔胡斯湾绝美的景色。构建景观的设计方法与融入自然的设计理念相互协调,将建筑所强烈展现的积极姿态与屋顶公园般的外观实用性紧密结合。

新西兰卡瓦卡瓦小镇上的毛利儿童保育及社区中心由Collingridge and Smith Architects建筑师事务所设计,项目特指为毛利部落儿童的早期教育所建造的一栋建筑。

象征符号的运用是设计的核心问题,例如人造假山景观呈豌豆状,象征着孕育万物的"地球母亲的子宫"。建筑所要表达的寓义是从地表

the interior explores the contrast between the exposed concrete roof, reflecting its structural reality, and the light and fine inner partitions and furniture, mainly of plywood. Energy efficiency is again provided by the earth as a natural insulator.

Focusing on the building landscapes examples, the Science Hills Komatsu, in Ishikawa, Japan, designed by UAO, is located on the site of a former factory, in an extensive and artificial urban context. The museum program includes a 3D theater, an event hall, multipurpose halls, an exhibition area, a science lab, seminar and workshop areas, lobbies and administration areas. Its image is dominated by the curving rooftops covered with grass, creating a series of artificial hills that visitors can clamber across.

In this case, dissimulation through the long curves is an affirmative gesture itself, although a restrained one. The grass roof compensates the lack of green areas in the surroundings, including a pathway for the public park. Its contemporary architectonic expression is dominated by the exposed concrete, the glass and the white painted steel. Unexpectedly, the project affirmative gesture is limited to the building itself, having no continuity in the waterproof large superficial car parking area, with no single tree.

The Moesgard Museum, in Aarhus, Denmark, designed by Henning Larsen Architects, is located in the hilly landscape of Skåde, surrounded by green areas. It is characterized by its sloping rectangular shaped roof of grass, which seems to grow out of the land, allowing for a park use during summer and a toboggan run during winter's snowfall. Its interior's museum program is composed by terraces, inspired in the archaeological excavations and in the layers of history.

Although partially dissimulated on the ground, the building intentionally assumes the idea of creating a new visual landmark on the territory, to what its dimension and the visible presence of the concrete contribute. The sloping roof-scape also creates a "natural" platform, offering an outstanding view of the Aarhus Bay. The building landscape approach goes together with nature integration, coherently combining the strong affirmative gesture with the exterior use of the park-roof.

The Maori Childcare and Community Center, in Kawakawa, New Zealand, designed by Collingridge and Smith Architects, specifically addresses the design of an early childhood building for a Maori tribe. Symbolisms become a central question, e.g., the artificial hill pres-

小松科学博物馆，日本石川
Science Hills Komatsu in Ishikawa, Japan

天然生长而出，但是除屋顶材料之外，建筑所呈现的姿态并没有表达所在区域的景观特点；整栋建筑位于地上，人行步道和停车场紧邻人造假山，这种建造方式削弱了假山的连续性。项目经过伪装处理，与其说是景观一体化建筑，倒不如说展现了积极的建筑姿态。入口一侧连续玻璃立面的设计，既与空间的内部设计相互呼应，又顺应了周边不利的环境特点。

Jorge Mealha设计的位于葡萄牙奥比多斯的科技园中央大楼，毗邻一个历史悠久的小镇，是位于乡村地区的一个企业孵化器。建筑为中空的四方体结构，悬浮在小面积空地之上，周围环绕着敏感的土地造型。

本案的设计有意强化地上的白色建筑与地下建筑之间的反差：前者体现出规则性、科技性，凌空跃出下方的地表平面，与地面几乎完全没有接触；后者则体现出不规则性和生态性，掩映在绿植地表下方，面向一个公共广场开放。公共广场的设计灵感源自于葡萄牙极具代表性的一种被称为"terreiros"的公共空间。本设计在创建一处新景观的同时，又呈现出一种概念上的双重姿态：(1) 建造绿色结构、地下建筑结构以及广场，使建筑场地融入周围环境；(2) 为景观增添新的积极元素——即纯白色的方形建筑物。显著的双重性揭示出项目内部极强的一致性，例如回字形的通路结构、建筑材料的选择——采用相同的半透明垂直多孔不锈钢板，建筑下方的钢板涂以风化色，上方为白色。

接下来，我们将以这七个项目为例详细说明建筑、区域和自然三个概念之间的关系，尤其侧重于项目对于重塑现有景观所发挥的重要作用。讨论主要围绕景观一体化与构建景观两种建筑理念之间的双重矛盾性展开。但最终的问题仍然是关于建筑质量的问题，我们只能等待未来的检验：这七个项目中哪一组建筑的姿态会经不起时间的考验而变得过时？哪一组能够一如既往地延续其内部一致性，始终保持刚竣工时的精工品质，且历久弥新？

ents a form of bean, signifying "the womb of earth mother", from where all life is born from. Having the intention of a building literally growing out of the land, the gesture hasn't such a territorial expression apart from the roof material; the building is placed above the ground and the artificial hill ends up immediately against the footpath and the car parking, weakening its continuity. It's a building with a camouflage, therefore much more an affirmative gesture than a landscape integration one. The opened side presents a glass continuous surface, answering to the internal distribution of spaces and to the passive environmental design features.

The Technological Park Central Building, in Óbidos, Portugal, designed by Jorge Mealha, is a business incubator unit in a rural area, close to the historic town. Located on a site with a small pending, it creates a square building with a hollow center, surrounded by a sensitive land modeling.

This case intentionally reinforce the contrast between the horizontal white square building, regular, technological, elevated and almost not touching the ground, and the underground building, irregular, organic, dissimulated under the green surface and opening to a public square which is inspired in the typical Portuguese public spaces, the "terreiros". It carries out a double conceptual gesture to build a new landscape, by creating: (1) the site integrated context, built through the green structure, the underground building and the square, and; (2) the affirmative element on the landscape, which is the pure white square building. This apparent double project reveals a strong internal coherence, e.g., in the circulation structure or in the materials, using similar translucent vertical perforated steel panels, weathered below and white painted above.

The seven projects presented in the next pages illustrate the conceptual relation between architecture, territory and nature, focusing particularly on how they contribute to shape the existing landscape. The discussion was centered on the ambivalence between the landscape integration and the creation of new landscapes. But a final question remains, regarding the quality of architecture, waiting for a future answer: which of these seven project gestures may become dated and which will maintain their internal coherence, surviving over the time with the same quality and freshness they present when they were just finished? João Pedro T. A. Costa

安蒂诺里酒庄

Archea Associati

建筑基地被独特的基安蒂山脉环绕,大面积的葡萄园覆盖其内,处于佛罗伦萨与锡耶纳的中间地带。客户具备深厚的文化背景,因此能够通过建筑来突出建筑景观和周围环境的完美融合,借以表达这片葡萄酒庄园的文化和社会价值。因此,功能成为整个设计过程中最为核心的部分,设计师的心思集中于设计这样一座建筑物,它能够给人带来的地貌体验,正是人们所想达到的与自然共生的最真实表达,并融人类文化、成就及其工作环境和自然环境于一体。酒庄的实体和精神构建以深植于这块土地的强烈联系为纽带,为使建筑形象隐藏起来,且能够与周边景观完美融合,这种关系特别紧密,也让人有些难熬(从经济投资的角度来看)。

该项目的目标是与建筑物以及乡间景观融合;由于屋顶被改造成了一片葡萄种植园,所以这座工业综合体从外观上成了景观的一部分。沿着轮廓线设置的两道水平开口将连绵的屋顶种植园打断,把日光引入建筑内部,同时以虚构立体模型的方式为屋内提供了观赏风景的视角。外立面采用建筑物典型的表达方式,沿着自然的坡度向水平方向延伸,间隔着一行行的葡萄藤,而这些葡萄藤与土地一起构成了"屋顶覆盖层"。

这些开孔或开口隐约地揭示出地下的室内设计:办公区布置得像地窖上的瞭望台,用于生产葡萄酒的区域沿较低层排列,而灌装区和储存区沿较高层布置。这座酒庄隐藏的核心区,也就是葡萄酒在木桶中酿成的地方,以其幽暗且韵律分明的赤陶拱顶传达出这个隐藏空间的神圣特征,不是因为不想让人们看到这个空间,而是为了保证这种葡萄酒的缓慢陈酿所需的理想温度与湿度条件。

细品这座建筑物的各个部分就可察觉到,其高度上的安排既遵循了葡萄酒生产流程向低处走的特性(就像受到重力作用一样),从到达点至发酵罐再到带有拱顶的地窖,还遵循着来访者逆行向上的次序,从停车场到酿酒区再到葡萄园,经过设置了压榨机的生产和展示区、餐后酒的陈酿区,并最终到达餐厅和布置着礼堂、博物馆、图书馆和品酒区以及销售区的楼层。

办公室、行政区和经理办公室位于高层,间隔着一系列的内部庭院,这些庭院由散布在葡萄园屋顶各处的圆形孔洞来提供自然照明。这个系统还用于为客房、管理员宿舍提供照明。项目所采用的材料和技术让人联想起当地崇尚简约的传统,并清晰合理地表达了经过精心安排的自然的主题,这些同时体现在赤陶的运用,以及聪明地运用土地天然产生的能量来冷却和保温这座酒庄的做法上,为葡萄酒的生产创造了理想的气候条件。

Antinori Winery

The site is surrounded by the unique hills of Chianti, covered with vineyards, half-way between Florence and Siena. A cultured customer has made it possible to pursue, through architecture, the enhancement of the landscape and the surroundings as expression of the cultural and social value of the place where wine is produced. The functional aspects have therefore become an essential part of a design itinerary which centers on the geomorphological experimentation of a building understood as the most authentic expression of a desired symbiosis and merger between anthropic culture, the work of man, his work environment and the natural environment. The physical and intellectual construction of the winery pivots on the profound and deep-rooted ties with the

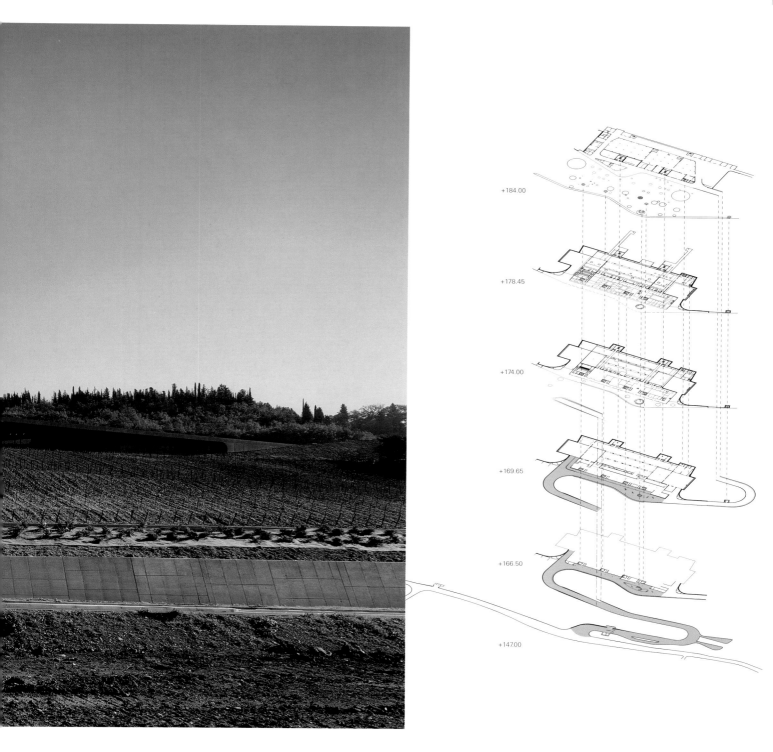

land, a relationship which is so intense and suffered (also in terms of economic investment) as to make the architectural image conceal itself and blend into it.

The purpose of the project has therefore been to merge the building and the rural landscape; the industrial complex appears to be a part of the latter, thanks to the roof, which has been turned into a plot of farmland cultivated with vines, interrupted, along the contour lines, by two horizontal cuts which let light into the interior and provide those inside the building with a view of the landscape through the imaginary construction of a diorama. The

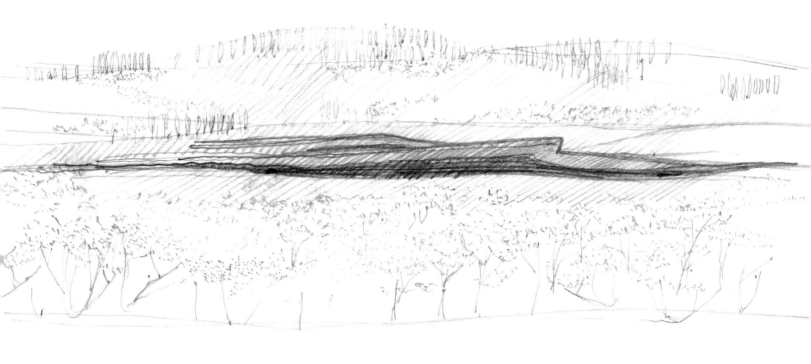
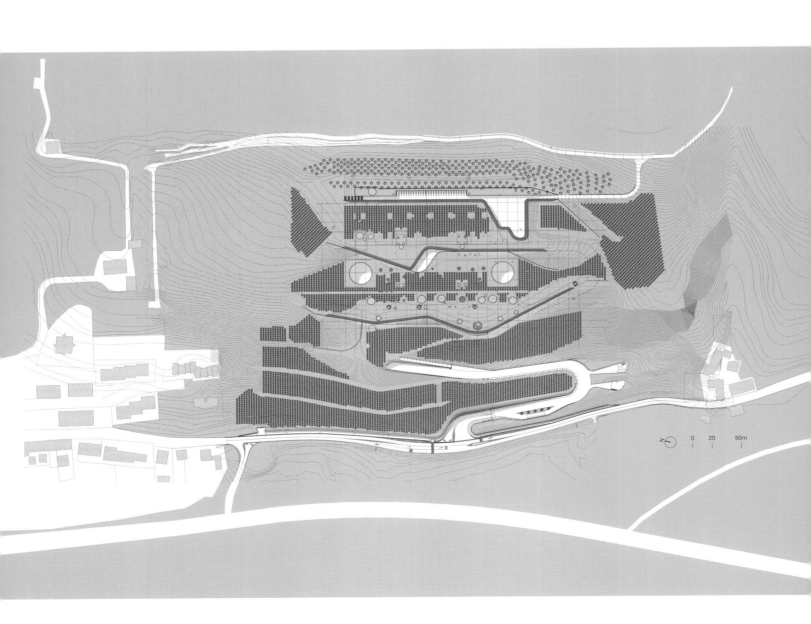

1 入口——门房	11 橡木酒桶酒窖	21 品酒室	31 餐后酒酒窖	
2 称重站	12 大桶酒窖	22 工作坊	32 油装瓶	
3 传动轴	13 葡萄榨汁室	23 档案室	33 仓库	
4 分离筛	14 露台	24 技术室	34 医务室	
5 游客大厅	15 接待处	25 拱顶上方	35 陈酿发酵区	
6 入口——电梯	16 办公室	26 自助餐厅	36 葡萄酒装瓶区	
7 瓶装酒酒窖停车场	17 礼堂	27 料斗	37 院子	
8 瓶装酒酒窖	18 博物馆	28 餐厅	38 橄榄园停车场	
9 储存区	19 商店	29 管理员宿舍		
10 橡木酒桶广场	20 等候室	30 葡萄风干室		

1. input - concierge	11. barrique cellar	21. tasting room	31. vinsanto cellar
2. weigh station	12. vat cellar	22. workshop	32. oil bottling
3. shaft	13. wine press room	23. archive	33. warehouse
4. bar screen	14. terrace	24. technical room	34. infirmary
5. visitors' hall	15. reception	25. above vault	35. bottle ageing
6. entrance - elevators	16. offices	26. cafeteria	36. wine bottling
7. bottle cellar parking lot	17. auditorium	27. hopper	37. yard
8. bottle cellar	18. museum	28. restaurant	38. olive orchard parking lot
9. house reserve	19. shop	29. keeper's house	
10. barrique square	20. waiting room	30. grape drying	

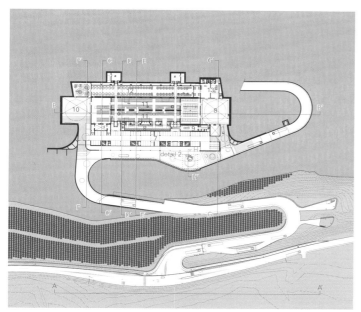

二层 second floor

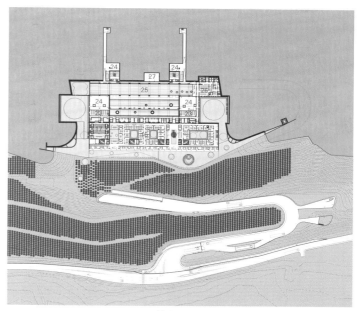

四层 fourth floor

五层 fifth floor

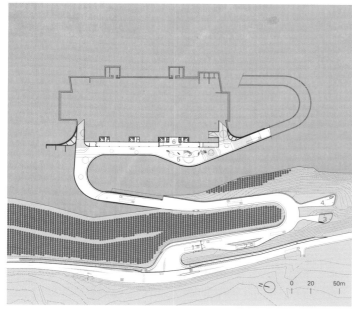

一层 first floor

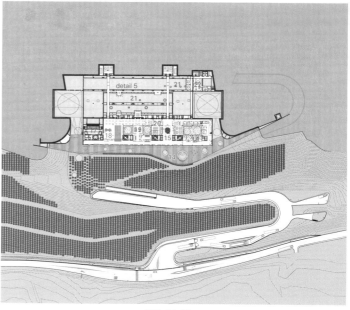

三层 third floor

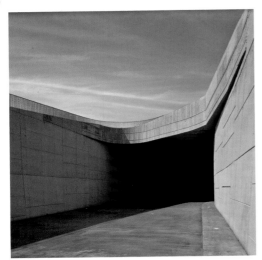

项目名称：Antinori Winery
地点：Bargino, San Casciano Val di Pesa, Firenze, Italy
建筑师：Laura Andreini, Marco Casamonti, Silvia Fabi, Giovanni Polazzi
艺术总监：Marco Casamonti / 工程：HYDEA
建筑现场监督：Paolo Giustiniani
结构设计：AEI Progetti
厂房设计：M&E Management & Engineering
酒庄顾问：Emex Engineering, Marchesi Antinori
总承包商：Inso
功能：winery, office museum, auditorium, restaurant, viability, manoeuvring and green area, depuration
用地面积：128,300m² / 总建筑面积：39,700m²
设计时间：2004 / 竣工时间：2012.10
摄影师：
©Pietro Savorelli (courtesy of the architect) - p.10~11, p.12~13, p.16[top-left, middle-right], p.17, p.18~19, p.20~21, p.25, p.26, 27[top]
©Leonardo Finotti (courtesy of the architect) - p.16[middle-left, bottom], p.23, p.24, p.27[bottom]

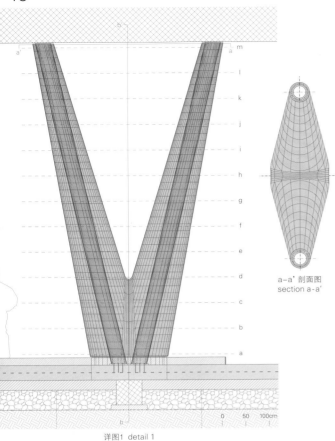

a-a' 剖面图
section a-a'

详图1 detail 1

1. sheet coating of corten steel calendered sp.3mm
2. filling cement based
3. fireproof plaster
4. steel structure made of tube 300, sp.40mm

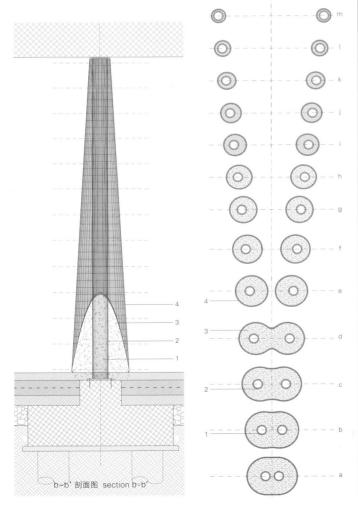

b-b' 剖面图 section b-b'

facade, to use an expression typical of buildings, therefore extends horizontally along the natural slope, paced by the rows of vines which, along with the earth, form its "roof cover".

The openings or cuts discreetly reveal the underground interior: the office areas, organized like a belvedere above the barricade, and the areas where the wine is produced are arranged along the lower, and the bottling and storage areas along the upper. The secluded heart of the winery, where the wine matures in barrels, conveys, with its darkness and the rhythmic sequence of the terracotta vaults, the sacral dimension of a space which is hidden, not because of any desire to keep it out of sight but to guarantee

the ideal thermo-hygrometric conditions for the slow maturing of the product.

A reading of the architectural section of the building reveals that the altimetrical arrangement follows both the production process of the grapes which descends (as if by gravity) – from the point of arrival, to the fermentation tanks to the underground barrel vault – and that of the visitors who on the contrary ascend from the parking area to the winery and the vineyards, through the production and display areas with the press, the area where vinsanto is aged, to finally reach the restaurant and the floor hosting the auditorium, the museum, the library, the wine tasting areas and the sales outlet.

The offices, the administrative areas and executive offices, located on the upper level, are paced by a sequence of internal court illuminated by circular holes scattered across the vineyard-roof. This system also serves to provide light for the guesthouse and the caretaker's dwelling. The materials and technologies evoke the local tradition with simplicity, coherently expressing the theme of studied naturalness, both in the use of terracotta and in the advisability of using the energy produced naturally by the earth to cool and insulate the winery, creating the ideal climatic conditions for the production of wine.

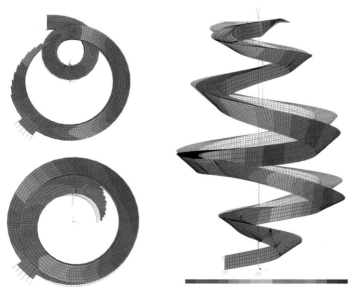

螺旋楼梯结构方案
spiral staircase structural schemes

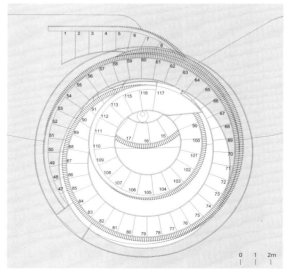

详图2 detail 2

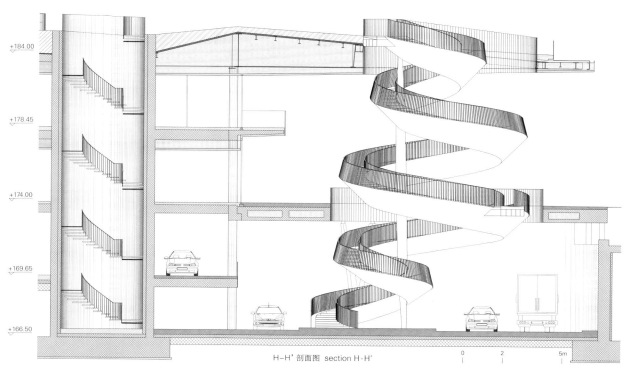

H-H'剖面图 section H-H'

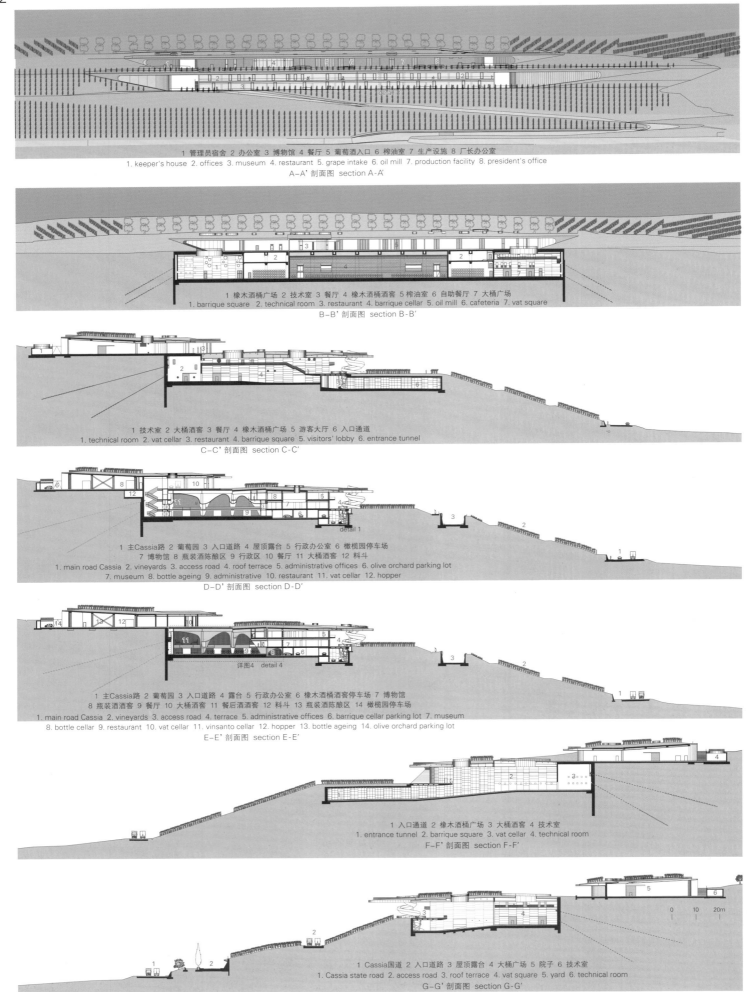

1. corrugated sheet metal with screed
2. HSU steel beams
3. suspended terracotta ceiling
4. glass door made with thermal break sections
5. rod railing
6. terracotta floor
7. reinforced concrete floor slab
8. insulating closure panel
9. metal framework supporting glazing
10. painted steel panel for built-in roller blinds
11. linear ventilation grid
12. double glazing panes
13. metal framework for supporting glazing
14. raised floor with carpet finish
15. reinforced concrete structure
16. metal framework for supporting glazing
17. floor slab made with a spiral prefabricated structure
18. corten steel plate
19. corten steel soffit cladding
20. linear ventilation grilles
21. suspended plaster ceiling

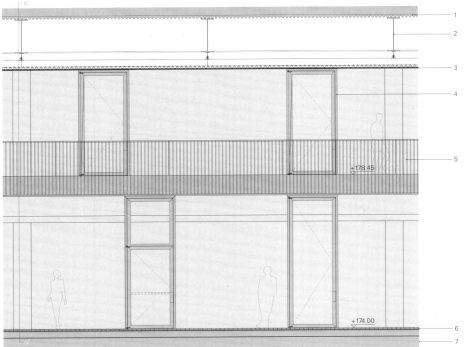

详图3 detail 3

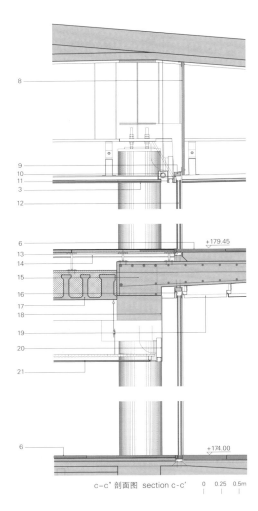

c-c' 剖面图 section c-c'

拱顶覆层发展剖面
vault cladding progression sections

详图4 detail 4

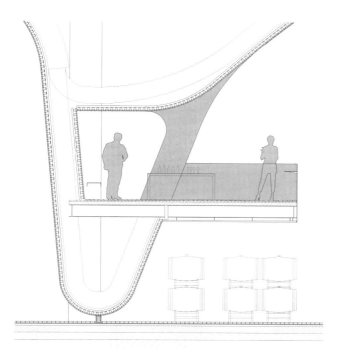

d-d' 剖面图
section d-d'

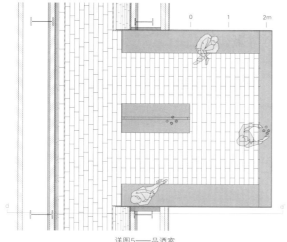

详图5——品酒室
detail 5 _ tasting room

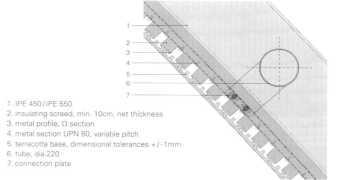

1. IPE 450/IPE 550
2. insulating screed, min. 10cm, net thickness
3. metal profile, Ω section
4. metal section UPN 80, variable pitch
5. terracotta base, dimensional tolerances +/-1mm
6. tube, dia.220
7. connection plate

赤陶拱顶细部
terracotta vault detail

1. vat cellar
2. passage for open tasting rooms
3. inspection walkway
4. barrique cellar
5. glazed tasting room
6. sorting corridor
7. offices
8. museum
9. bottle cellar

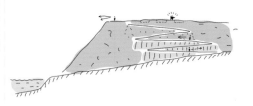

Vučedol考古博物馆

Radionica Arhitekture

  Vučedol位于多瑙河右岸,武科瓦尔市中心下游4.5公里处。以它的名字命名的文化——Vučedol文化与美索不达米亚的苏美尔文明处于同一时代。美索不达米亚是埃及古国、特洛伊城的发端。Vučedol文化建立初期位于Srijem和东斯拉沃尼亚地区,经过发展最终遍布整个克罗地亚以及东欧和中欧11个国家的部分地区。

  在这个地方的首次考古工作发现促成了19世纪末期的Streim家族的建立,他们是Vučedol遗址的拥有者,其经过重建与改造的房屋仍然矗立在考古公园中。这一发现极大地鼓舞了考古研究工作的进一步开展,其中最为著名的是德国考古学家罗伯特·R.施密特于1938年发掘出的正厅和举世闻名的Vučedol陶鸽。

  格拉达茨曾是一个铸造中心,由于在铸造过程中会释放出有毒气体,所以将格拉达茨作为村庄的卫城和宗教中心,与村庄分隔开来。Vučedol的房屋有许多圆柱形穴室,用作储藏室,有时也用作献祭的动物和人类的墓穴。Vučedol建筑的特点是房屋有圆角,以柳条黏土筑墙,在房屋中心设一个大壁炉。

  对Vučedol文化的系统研究始于1984年。在考古发掘之初,在遗址之上建立一座考古博物馆的计划就已经确立。20世纪90年代初期,考古研究工作因战乱被迫停止,在2000年场地的地雷被清除之后才恢复。

  考古遗址占地超过6英亩(约2.43ha),大部分被树木和葡萄园所覆盖。地形呈上升趋势,从博物馆入口处的平均海拔90m上升至格拉达茨遗址所在地的平均海拔大约110m。园区内的主要建筑有:博物馆、修复还原的Streim家族住宅(作为考古研究中心)、古代手工艺作坊(用于展示Vučedol文化工艺品)、格拉达茨旧址的地下建筑和重建于该建筑之上的正厅。

  建造博物馆的基本理念是使建筑与地形相契合,为此,建筑设计将博物馆的绝大部分主体结构埋入地下,只有建筑的立面向景观开放。整栋建筑外形遵循地势,呈蜿蜒盘曲的蛇形。从绿色植被覆盖的屋顶即可到达博物馆的考古遗址。除蛇状外形之外,建筑材料的选择也体现出将博物馆与地形相融合的设计理念。建筑外墙面为红砖层,这是最接近现场风格的自然材料。

  博物馆内部被划分为不同的功能区。一层是包括咖啡厅、更衣室等专为游客服务的空间。办公室及储藏室一半设置在地下,由一层进入。其余的室内展览空间被分成若干层次,相互之间以坡道相连。每层展区均设置出口,方便游客随时离开博物馆,到室外的建筑顶部继续游览。由于建筑主体结构多半掩埋于地下,且空间纵深体量较大,因此设计了中庭以增加室内空间的采光。

  博物馆整体采用钢筋混凝土结构,其中包括基础底板层、蛇形曲线之间的纵向墙、横向的梯形墙、纵横交错的支撑梁和屋顶。内部处理极为简单,直接采用裸露的混凝土涂以黑漆作为墙体,地面铺设了浅色橡木地板。

  园区未来将增加旅游和娱乐活动类规划,还将增设大型停车场和游船停靠码头。

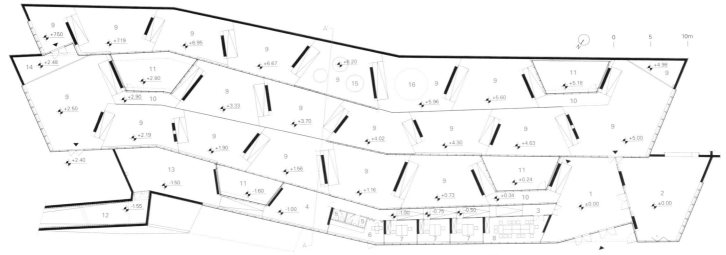

1. 大厅 2. 咖啡馆 3. 走廊/坡道 4. 作坊 5. WC 6. 办公室厨房 7. 办公室 8. 会议室 9. 展厅
10. 高地 11. 中庭 12. 仓库入口 13. 仓库 14. 水族箱 15. 考古发现 16. 天文台
1. hall 2. cafe 3. corridor/ramp 4. workshop 5. WC 6. office kitchen 7. office 8. meeting 9. exhibition
10. plateau 11. atrium 12. depot access 13. depot 14. aquarium 15. archaeological finds 16. planetarium

Vučedol Archaeological Museum

Vučedol is located on the right bank of the Danube River, 4.5km downstream from the center of Vukovar. Culture that bears its name, the Vučedol culture, is contemporary with Sumerian period in Mesopotamia, the Old Kingdom of Egypt and the beginnings of Troy. In its initial phase it was located in Srijem and Eastern Slavonia, and in the late stage, it was spread over the whole of Croatia and parts of 11 countries in Central and Eastern Europe.

First archaeological findings, on the site, made the family Streim in the late 19th century, who was the owner of Vučedol site and whose reconstructed house still stands in the archaeological park. This prompted further research, of which the most notable was the one made in 1938, when a German archaeologist, Robert R. Schmidt excavated Megaron and found the world-famous Vučedol Dove.

All of Gradac was a casting center, and because of casting and the poisonous gases released from that process, Gradac was separated from the rest of the village, and was as a kind of acropolis and religious center of the village. The Vučedol houses had numerous cylindrical tombs that were used for storage room, and sometimes as the tombs for ritually sacrificed animals and humans. Vučedol houses are characterized by rounded corners and walls of wicker, coated with clay, with a large fireplace in the center.

Systematic research of Vučedol began in 1984 and already at the beginning of the excavations, there was the idea of founding an archeological museum on the site of the excavations. The

项目名称：Museum of Vučedol Culture
地点：Vukovar, Croatia
建筑师：Iva Peji, Goran Rako, Josip Saboli, Mario karijot
展览设计：Vanja Ilić
施工工程师：Radionica statike
安装：Arhingtrade d.o.o., KGH d.o.o.
承包商：Ing-Grad d.o.o., Monte-Mont d.o.o.
用地面积：82,800m² / 建筑面积：2,400m² / 总建筑面积：2,500m²
设计时间：2008—2011 / 施工时间：2009—2013
摄影师：©Boris Cvjetanović (courtesy of the architect)

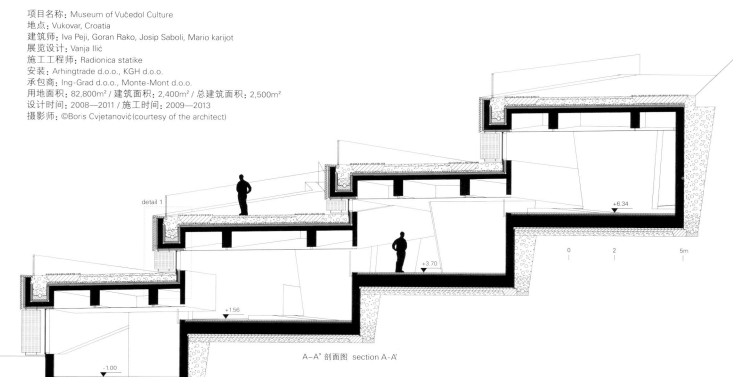

A-A' 剖面图 section A-A'

research stopped in the early 1990's because of the war and resumed in 2000 after the de-mining of the area.

The archaeological site extends to just over 6 acres and is mostly covered with woods and vineyards. The terrain rises from about +90m AMSL where the entrance to the museum is, to approximately 110m AMSL where the Gradac site is located. The main buildings in the park are: the museum, restored Villa Streim(which will serve as a research center for the archaeologists), old crafts workshops(for demonstrations of crafts from Vučedol culture), underground building on the site of Gradac and reconstructed building of Megaron above it.

The basic idea behind the concept of the museum was integration into the terrain, which is achieved with the museum design which is mostly buried in the ground and only the facade is open to the landscape. Its shape, as serpentine, follows terrain, and on whose green roof you can reach the archaeological sites over the museum. Integration into the terrain is achieved, except with serpentine form, by selecting materials, and so was brick selected for the outer coating, a natural material that most resembles the ground at the site.

The interior of the museum is divided into different sections. Ground floor contents, such as coffee shop and dressing rooms, are designed for visitors. Offices and storage are accessible from the ground and are partially placed in the basement. The rest of the interior exhibition space is divided into several levels, interconnected with ramps. On separate levels of exhibition space it's possible to exit the museum, and continue touring on the roof surface. Due to the fact that the building is mostly buried, but also because of the depth of space, atriums were made to further illuminate the interior space.

The entire structure of the museum is made of reinforced concrete. It consists of a foundation slab and the longitudinal walls between each serpentine, transverse trapezoidal walls, longitudinal and transverse beams and roof slabs. Treatment of the interior is simple; the walls were left as bare concrete, finally painted in black, and on the floor light oak was placed.

As part of the future park, tourist and recreational activities are planned, as well as a large parking lot for cars and buses, and a pier for tourist boats.

1. suspended ceiling
    - concrete slab 20cm   20cm
    - PVC waterproof membrane   0.2cm
    - extruded polystyrene insulation   10cm
    - geotextile
    - draining element   6cm
    - coarse gravel
    - geotextile mat
    - soil   30cm
2. coarse gravel
    - dimpled sheet membrane   0.5cm
    - extruded polystyrene insulation   10cm
    - PVC waterproof membrane   0.5cm
    - concrete wall   20cm
    - thermal insulation   10cm
    - ventilation   5cm
    - facade brick   12cm
3. suspended ceiling
    - concrete slab   20cm
    - impact sound insulation   2cm
    - separating layer   0.02cm
    - screed   6cm
    - parquet   2cm
4. coarse gravel
    - dimpled sheet membrane   0.5cm
    - extruded polystyrene insulation   10+5cm
    - PVC waterproof membrane   0.2cm
    - concrete wall   3cm

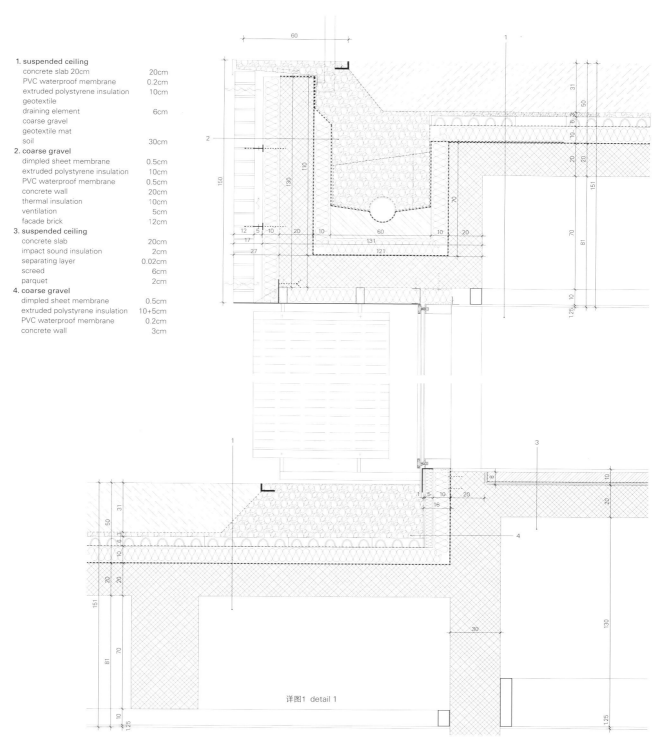

详图1 detail 1

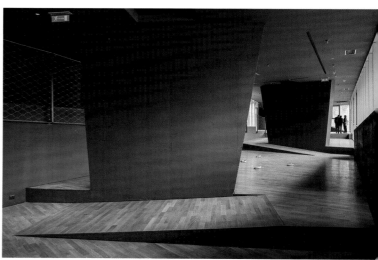

克尔科诺谢山环境教育中心

Petr Hájek Architekti

由克尔科诺谢山国家公园 (KRNAP) 建造并负责管理的克尔科诺谢山环境教育中心 (KCEV) 展现了一种全新的教育机构形象。KRNAP建造KCEV的初衷是广泛推行他们的公众教育活动。

位置

从弗尔赫拉比市远观，城市的重心并不在主广场内，而是位于城堡周围美丽的公园中。新建的教育中心正是坐落在城堡对面的公园内。

建筑

项目的主要设计理念是建筑本身既可作为一栋独立的建筑物，也可充当研究克尔科诺谢山脉地形地貌的极好教材。本案是建筑与景观合而为一的典范，建筑的几何形状设计灵感来源于克尔科诺谢山脉起伏绵延的几何线条——每个坡度都能在真实的自然地形中找到对应。屋顶呈现的几何效果并不是随意做一些简单的切削处理，而是充分考虑了克尔科诺谢山的山形构造。建筑外观铺设绿色植被，宛如"山地草场"，内部天花板由象征岩石的混凝土面板构成。衔接各个屋顶平面的屋脊既是显著的地形符号，又与周围的山体景观形成呼应。

现有建筑与新建的教育中心之间设置了一整面玻璃墙，人们可以透过它看到建筑内部，间接参与室内的讲座、研讨和授课。这种设计进一步强化了空间的公共氛围。建筑采用节能设计，使用热泵供暖，将热能损耗降至最低。建筑的主要承重结构是光滑的裸露混凝土，内部隔断由木板制成。家具根据胶合板的规格尺寸设计，避免材料浪费。

功能

建筑内部设有一个报告厅、一个实验室、一个图书馆、一个教室和一个展览厅，另有几间技术室、设备间和储藏室。建筑与行政大楼之间通过地下通道相连。

礼堂可作报告厅和科学会议室使用，由于配备了宽屏幕布和立体声音响系统，还兼具小型电影放映厅的功能。为适应活动需要，礼堂还可作为扩展的展厅、实验室以及教室使用。

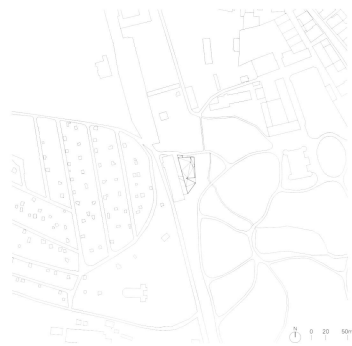

1. Obří Hřeben, way up to Sněžka (1,602m) in direction to Svorova Hora (1,411m)
   slope of the vector: 16°
3. Pod Černohorským Rašeliništěm, way down to Černá Hora across Vlašské Boudy
   slope of the vector: 4°
7. Sněžné Jámy, valley glacier on the polish side close to the chalet of Sněžné Jámy (1,490m)
   slope of the vector: 44°
9. Szrenica, way up to the polish side between Szrenica (1,361m) and Koňské Hlavy (1,297m)
   slope of the vector: 7°
11. Růžová Hora, way down from Růžová Hora (1,390m) across Lví Důl
    slope of the vector: 3°
12. Krakonošova Zahrádka, Úpská Jáma over Studniční hora (1,554m)
    slope of the vector: 32°

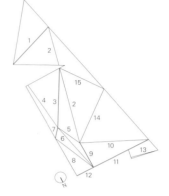

教学过程中,学生与实践教学和理论教学的初次接触至关重要,因此设计将实验室与教室、图书馆紧密相连。

教室的开放空间与实验室、图书馆相通,学生的学习生活多半在这里进行。教室还可改装成绘画室、多媒体工作室或阅览室,非常便易。

KCEV的设计目标是建造一个环境教育基地,满足组织公开讲座、国际会议和项目研究等功能需求,因此在车库兼画廊空间的设计方面不做过多规划,以便针对上述需求自由调整内部空间的使用。同时这也成为对学生教育的一个重点,即通过这种设计深化学生对自然价值的认同与尊重。

Krkonoše Mountains Environment Education Center

Krkonoše Mountains Environment Education Center (KCEV) presents a new education institution established and administrated by the management of the Krkonoše Mountains National Park (KRNAP). The intention of KRNAP was to significantly extend their educational activities concerning the public.

Location

Seen from a distance in the city Vrchlabi, it doesn't have its center of gravity within the main square, but rather in the beautiful park surrounding the castle. The site for the new center was set in the park right opposite the castle.

Architecture

The main idea was that the building in its principle could be seen as an object or doctrine to study the topography of the Krkonoše Mountains. The object is a hybrid between a building and the landscape. Its geometry refers to the geometry of the Krkonoše Mountains – each single slope has its natural counterpart. The roof is not some minimized cut-out, but contains information about the allover formation of the Krkonoše Mountains. The exterior is designed like a "mountain meadow", whereas the inner ceiling consists of face concrete, which symbolizes the rocks. The edges connecting the single parts of the roof are highlighted by a graphic symbol and the description of the corresponding mountain.

The ambience of the public space between the existing building and the new center, is emphasized by a glazed wall, which

项目名称：Krkonoše Mountains Center for Environmental Education
地点：Vrchlabi, Czech Republic
建筑师：Petr Hájek Architekti
概念设计参与者：Helena Línová, Michal Volf
设计参与者：Cornelia Klien, Andrea Kubná, Ondřej Lipenský, Helena Línová, Martin Prokš, Martin Stoss, Michal Volf, Jan Kolář
平面设计师：Kristina Ambrozová
土木工程师：Jan Kolář / Geometry of the shell: Jaroslav Hulín
便携式家具设计：Cornelia Klien, Martin Stoss
委托人：Krkonoše Mountains National Park Administration
用地面积：1110m² / 建筑面积：792m² / 有效楼层面积：962m²
设计时间：2009 / 竣工时间：2011—2014
摄影师：©Benedikt Markel (courtesy of the architect)

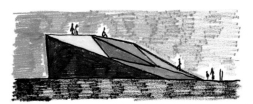

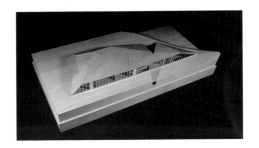

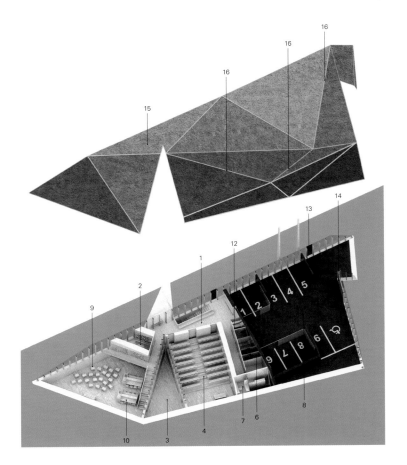

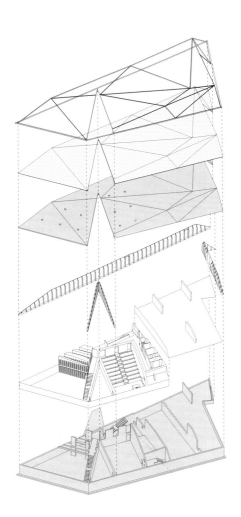

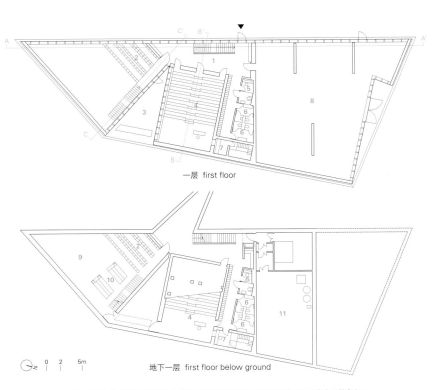

一层 first floor

地下一层 first floor below ground

1 入口大厅 2 图书馆 3 展览空间 4 礼堂 5 控制室/警卫室 6 卫生间 7 技术室 8 车库 9 教育大厅 10 实验室 11 技术设备室 12 门房 13 自行车入口 14 车道 15 景天属植物绿色屋顶 16 克尔科诺谢山坡度

1. entrance hall 2. library 3. exhibition 4. auditory 5. control room/gatehouse 6. toilets 7. technicians cabin 8. garage 9. educational hall 10. laboratory 11. technical facilities 12. porter's lodge 13. cyclists entrance 14. driveway 15. green roof with sedum 16. vectors of the Krkonoše mountains

conveys the inner life to pedestrians and provides indirect participation in lectures, seminars and tuitions. The building itself is energy efficient with a minimum of heat loss and is heated with a heat pump. The main load-bearing structure is made of smooth, exposed concrete, whereas inner partitions consist of timber. The design of furniture is adapted for the size of plywood plates to prevent from wasted residues.

Program

The building contains a lecture hall, a laboratory, a library, a classroom, an exhibition hall, technical rooms, facilities and storages and is connected underground to the administration building.

The auditorium serves as a lecture hall as well as for scientific conferences. Due to a wide screen canvas and a stereo sound system it achieves the parameters of a small cinema hall. In order to extend activities, the auditorium can be connected with the exhibition space, the laboratory and the classroom.

The first contact with theoretical and practical teaching plays a central role in education, therefore the laboratory is combined with the classroom and the library.

In the open space of the classroom, which is combined with the laboratory and the library, students will spend most of their time. The classroom can easily be changed into a hall for drawing classes, a multimedia studio or a reading room.

The Garage/Gallery is designed as a hybrid. Due to its free plan the space may be adapted according to the requirements. KCEV was intended to become a place for environmental education, and should serve the organization of public lectures, international conferences and research projects. At the same time it will be a center for educating students, who will thus gain deeper awareness and respect for natural values.

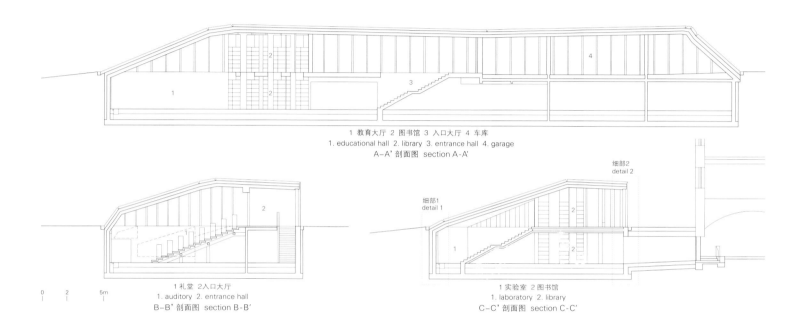

1 教育大厅 2 图书馆 3 入口大厅 4 车库
1. educational hall 2. library 3. entrance hall 4. garage
A-A' 剖面图 section A-A'

1 礼堂 2 入口大厅
1. auditory 2. entrance hall
B-B' 剖面图 section B-B'

1 实验室 2 图书馆
1. laboratory 2. library
C-C' 剖面图 section C-C'

几何形状解决方案
geometry solution

修剪前壳体形状
shape of the shell before cropping

选定的所有矢量线条的聚焦点设在地面下30m处
Selected focus point of all of the vectors was placed 30 meters under the ground

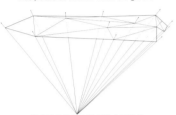

混凝土的聚焦点通过多方面设定
Concrete focus point was selected from various possibilities

多种形状在一点连接，但如果它们超过相同高度，就无法连在一起
Several shapes are joined in one single point, but if they are extruded of the same height, they cannot connect

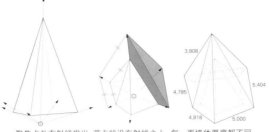

聚焦点处有射线发出，节点就设在射线之上，每一面墙体厚度都不同
There are rays from the focus point and nodes are placed on the rays, each single wall has its own thickness

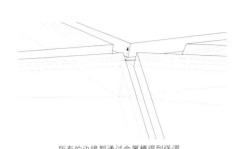

所有的边缘都通过金属槽得到强调，金属槽的几何形状取决于整个屋顶的聚焦点
All of the edges are emphasized by the metal trough, the geometry of which depends on the focus point of the whole roof

1. lightweight extensive green roof substrate, 80mm
2. roof with a setup of extensive sedum
   pre-cultivated mat, mixture of xerophilous undemanding plants – on a non-putrescent mat, 15mm
   lightweight extensive green roof substrate, 80mm
   anti-slip system – compounded grid, beams and sills
   structured water-storing textile, 8mm
   separating and protective mat, 5mm
   waterproofing layer – dual system of control and activation functions
   thermal insulation, 0.21 mpa with 10% linear deformation, bulk density 30-35 kg/m², l=0.034 w/mk 220mm – connected to the substrate mat
   vapour barrier e.g. 1x modified asphalt strip, punctually submitted to the substrate mat 4mm, first coat
   load-bearing ceiling, 260mm
3. l profile – supporting grid for the green-roof
4. drainage channel
5. concrete base for the slope channel
6. wood based wall cladding panel, face layer – multiplex
   wooden substructure tk. 30mm or 50mm, anchored to the concrete walls
   reinforced concrete wall, 300mm
   waterproofing layer against radon load – reference layer dualtek, 10mm
   2x extruded polystyrene, compression 200 mpa l= 0.038w/mk, 80mm+80mm
   protective lining with concrete bricks tk. 75mm
   with columns 450/450mm a 2m
   compacted embankment
7. wooden top layer certified for use
   as the top layer of the poor, laminated plywood – multiplex, adhesive, 20mm,
   dry bound, machine-finished concrete with plasticizer c25/30 with 2x embedded reinforcement net from ribbed build-ups, ø4/ø4-100/100, 90mm
   separating layer – pe foil tk. 0.2mm, overlapping with taped joints
   included sound installation for hard floors
   terminating strips, 40mm
8. expanded polystyrene eps 200, 80mm
   concrete slap with reinforcement 2x network 8x150x150, concrete c30/37, steel 10505 (r), 150mm
   form work of galvanized trapezoidal sheets tk. 1mm height 30mm
   compacted gravel backfill-dried moisture content below 2%, 500mm, reinforced concrete tub, 400mm
   protective concrete screed c25/30, 50mm
   waterproofing layer against radon load, dual system of control and activation functions
   base concrete tk. 150mm c30/37, 150mm
   gravel substrate 100mm, soil
9. convectors covered by an atypical grid
10. led-strip covered by acrylic glass
11. trapezoidal shaped glass facade, 54,600mm/4,140mm
    load-bearing columns, axial distance 1,200 mm
    fire glazing ei30
    the glass is towards the roof gradually covered with a translucent point print
    two integrated doors and two integrated ventilation diffusers in the facade
12. laminated plywood – multiplex, adhesive, 20mm
    dry bound, machine-finished concrete with plasticizer, 2x embedded reinforcement net, 80mm
    separating layer – pe foil, 0.2mm
    sound insulation for hard floors, 50mm
13. protection against moisture embedded in the ceiling construction, pe foil, 0.2mm
    reinforced concrete, 200mm
    plywood – multiplex, suspended on a steel grate, 30mm + 20mm
14. smoothed concrete, 120mm
    gravel bed, tk. 100mm
    gravel drainage layer
    protective layer of extruded polystyrene specified for inverted roofs, 100mm
    waterproofing layer against radon load, dual system of control and activation functions
    thermal insulation, 100mm
    vapour barrier, modified asphalt strip, 4mm
    concrete ceiling, 220mm

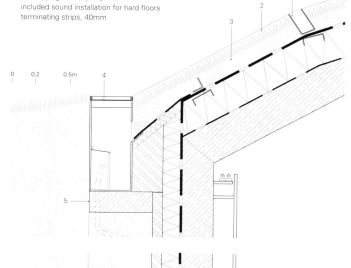

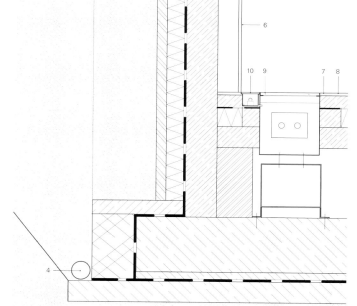

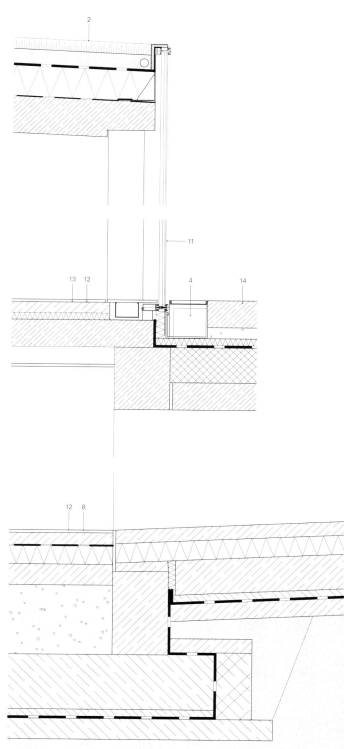

细部1 detail 1

细部2 detail 2

景观与建筑 Landscaping and Building

毛利儿童保育及社区中心
Collingridge and Smith Architects

这一专为毛利部落儿童设计的早期保育建筑位于新西兰的卡瓦卡瓦。该建筑不仅为孩子们提供了居所，而且孩子们在每天的生活中也受到了关于毛利文化和习俗的教育。同时建筑本身对环境产生的影响也被降至最低。

我们的设计理念基于毛利人的传统，在他们看来，所有的人类都是由地球母亲在海洋中的子宫所孕育的。同时，"陆地"一词在毛利语中也意味着"胎盘"。毛利部落的建筑历来以富于象征主义而著称，因此该设计是这样构思的：将土地塑造成子宫状，将建筑设计为其中的婴儿，这样一来，建筑就好像是破土而出的一样。

建筑物唯一的入口沿北立面设置，看起来就像是土地上的一个切口。这个切口象征着剖腹产术，也正是因为这样，所有的部落都有了属于自己的血统：他们的祖先Hine ā Maru是第一位因成功剖腹产而被记入史册的女性，600多年前，她通过剖腹产诞下了一个孩子，并且存活了下来。正是通过这个切口，孩子们象征性地进入了所谓的"光之世界"，他们可以在这里尽情地嬉戏。

这一建筑建立在沼泽地上，并以子宫般的岛屿形式呈现出来。它紧紧围绕着"所有生命都是由海洋孕育的"这一传统。人们修建了一座通往岛屿的桥。桥被设计成了部落的独木舟的形状，代表着部落祖先从哈瓦基到新西兰的征程。

堆在建筑物上方的泥土参考了毛利族的风格，而地表以下的内部再现了附近的萤火虫洞的形象，这是他们祖先被埋葬的地方，以及他们的防御工事建筑，他们的祖先卡维提非常聪明地利用地下庇护所来抵御外敌。建筑的圆形设计也是从传统城市建筑中得到的灵感。学会将被动的环境设计融入建筑中，这一点同样很重要。因此，所有看似具有"象征意义的"特点都有许多环境效用：所有的玻璃窗都面朝北就是为了最大程度地吸收太阳能。同时超保温的屋顶是为了最大限度地减少热量的散发，而位于南边，未安装暖气的流线空间也起到了一定的辅助作用。为了让内部更加舒适，裸露的混凝土结构和自然通风设施都使建筑在夏天保持凉爽，太阳能热水地热系统使建筑在冬天保持温暖，同时又尽量少消耗能源。

所有的空间借助的都是自然光线。在白天，建筑不需要借助电灯照明。污水都是现场进行处理的，干净、饱含营养物质的水用来浇灌经过绿化的屋顶。该建筑已提交"绿色之星"环境评级的评定，并有望获得六星认证。

Maori Childcare and Community Center

This design is an early childhood building for a Maori tribe(Ngāti Hine) in Kawakawa, New Zealand. The brief called for a building which would not only accommodate the tamariki(children) but teach them about their culture and customs on a daily basis whilst having a minimal impact on the environment.

Our concept for the building is based on the Maori tradition that all life is born from the womb of Papatūānuku(earth mother), under the sea: the word for land(whenua) in Maori also means placenta. Maori architecture is historically rich in symbolism, so the design is conceived by shaping the land into a womb-like form, with the building forming just like a baby within: the building literally grows out of the land.

The only opening to the building is along the north facade, and reads as a cut in the earth. This cut symbolically represents the

被动制冷与通风 passive cooling and ventilation

被动供暖与通风 passive heating and ventilation

雨水收集和净化系统 rainwater harvesting and purification system

南立面 south elevation

东立面 east elevation

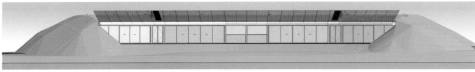
北立面 north elevation

西立面 west elevation

caesarian birth through which all of the iwi clients(tribe) take their lineage: their ancestor Hine ā Maru was the first recorded Maori woman to deliver a child by caesarian section and survive the procedure about 600 years ago. It is from this opening that the children symbolically enter the "world of light", where they play.

The building is located on marshy ground, with the "womb-like form" appearing as an island, relating back to the tradition that all land is born from under the sea. A bridge is formed to give access to the island, which is symbolically shaped into the tribal waka(canoe) representing the journey of the tribe forefathers from Hawaiki to Aotearoa(NZ).

The earth that mounds up over the building makes reference to Ngāti Hine. The interior, below the earth, represents the nearby Waiomio caves where the ancestors lay buried and the Ruapekapeka urban buildings (fortification) where the ancestor Kawiti cleverly used underground shelters as defense from attack. The circular form of the design also draws inspiration from traditional urban buildings. It was equally important to integrate passive environmental design features into the building, so all "symbolic" features have many environmental purposes: all glazing is oriented to the north for maximum solar gain, whilst the super-insulated earth roof results in minimal heat loss, which is further assisted by the unheated circulation space placed to the south. For further internal comfort, exposed concrete construction and natural ventilation allow the building to be cooled in summer, with minimal heating back-up in winter provided by a solar hot water underfloor system.

All spaces are naturally daylight and will need no additional electrical lighting during the daytime. All black water is treated on site and the clean nutrient rich water is used to irrigate the green roof. The building has been submitted for a Green Star rating and is anticipated to achieve six stars. Collingridge and Smith Architects

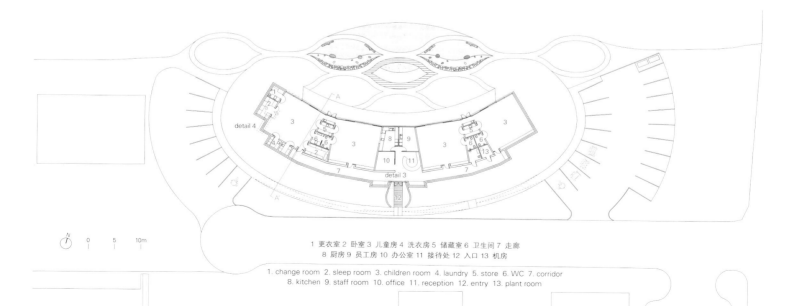

项目名称：Maori Childcare and Community Center
地点：Kawakawa, Northland, New Zealand
建筑师：Collingridge and Smith Architects(CASA)
项目建筑师：Phil Smith
设计团队：Graham Collingridge, Grayson Wanda, Chloe Pratt
结构工程师：McNaughton Consulting Engineers
机电工程师：Eco Design Consultants, WSP
预算师：Kwanto Ltd.
生产计划主管：Simon Yates
照明工程师：Lighthouse Remuera
景观设计师：Phil Smith, Jonathon Fulton Landscaping
甲方：Ngati Hine Health Trust
总承包商：Howard Harnett Builders
用地面积：3,965m² / 建筑面积：1,362m² / 有效楼层面积：572m²
造价：USD 2,200,000
施工时间：2011 / 竣工时间：2012
摄影师：©Simon Devitt(courtesy of the architect)

1 更衣室 2 卧室 3 儿童房 4 洗衣房 5 储藏室 6 卫生间 7 走廊
8 厨房 9 员工房 10 办公室 11 接待处 12 入口 13 机房

1. change room 2. sleep room 3. children room 4. laundry 5. store 6. WC 7. corridor
8. kitchen 9. staff room 10. office 11. reception 12. entry 13. plant room

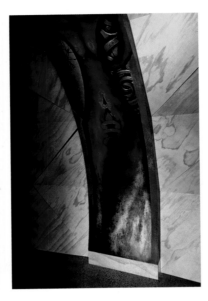

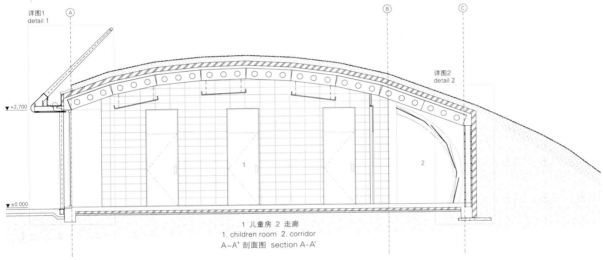

1 儿童房 2 走廊
1. children room 2. corridor
A-A' 剖面图 section A-A'

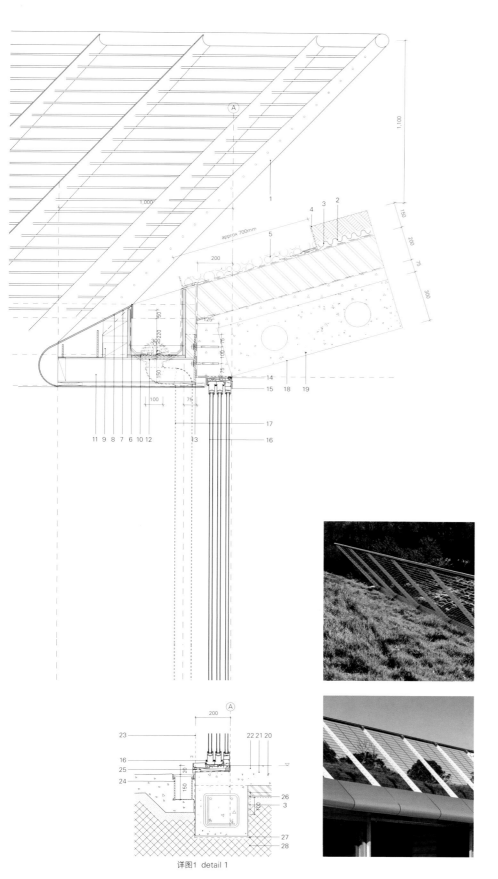

1. balustrade
2. green roof
3. waterproofing membrane
4. perforated angle glued to roofing membrane
5. pebbles to edge of green roof as per equus spec.
6. fascia downturn over plywood gutter
7. fascia
8. neatly seal between balustrade and alucobond junction with structural sealant. refer to alucobond specifications
9. H3.1 timber blocking (FSC certified)
10. gutter
11. 150UC 23 outriggers as per engineer
12. all-proof 168.100DRCR 100 dia domed roof clamp ring drain
13. gutter supports bolt fixed through insitu poured edge
14. compressible foam strip air seal
15. 25x25 anodised aluminium cover angle to match joinery fixed to u/s concrete and over joinery
16. external aluminium doors/windows
17. downpipes
18. u/s of dycore panels to be painted to resene ECNZ spec
19. roof structure
20. floor structure

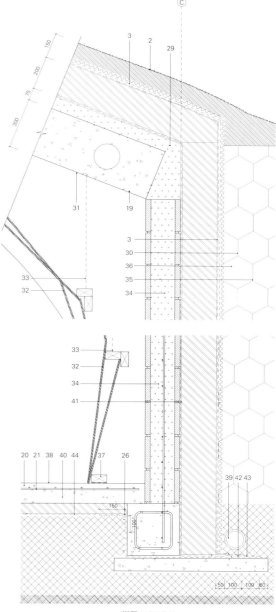

详图1 detail 1

详图2 detail 2

21. all steel reinforcing to footings and slab junctions to engineers design
22. fill remaining gaps
23. line of blockwork beyond
24. ACO KlassikDrain KS100 with perforated stainless steel grate installed as per manufacturers instructions
25. waterproofing layer - debobase 3.5 cs/f
26. reinforcement strip - debovix 3t/f p180
27. waterproofing layer - debobase 3.5 cs/f
28. compacted hardfill
29. insitu poured concrete to perimeter of dycore panels
30. polyrock
31. u/s of dycore panels to remain unlined
32. triangular interior plywood lining
33. proprietary hanging system for fixing plywood lining
34. concrete block structural walls
35. earth bank
36. gravel drainage layer to engineers specification
h3.1 90 x 45 fsc cert timber wall plate to fix plywood - fix over
37. bituminous dpm
38. floor finish
39. draincoil in filter sock
40. 75mm in-site lean concrete
41. bituminous primer - equus duo primer
42. vapour barrier - deboflex 2.5 t/f 175
43. cap sheet - duo ht 4 wgg/f c180 landscape
44. waterproofing layer - debovix 3t/f p180

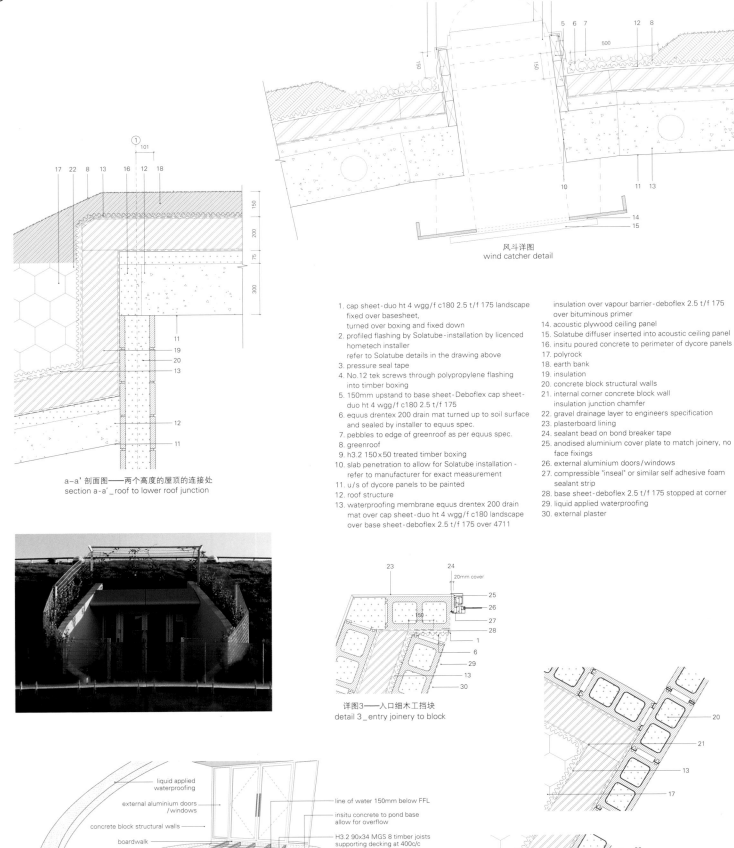

莫斯格史前博物馆
Henning Larsen Architects

新的莫斯格史前博物馆独一无二地坐落在丹麦奥胡斯Skåde丘陵景观当中。该建筑有一个巨大的斜坡屋顶，屋顶平台上面是草坪、青苔和各种色彩鲜艳的野花，这让建筑即便从海面上看过来也是一个绝对吸引眼球的地标。

矩形的屋顶平台就像是从景观中生长出来的一样。夏天里，在这个屋顶草坪上可以进行野餐、烧烤，举办户外讲座和传统的仲夏节篝火晚会。而当冬雪到来，斜坡屋顶又摇身一变成了当地最好的雪橇乐园。

建筑的室内设计构想是要获得一种变化的阶梯状台地景观，其灵感来源于考古发掘工作总是通过挖掘历史来层层揭示遗失的文明的特点。参观者可以在一系列生动的展品和科学实验中穿行——像一个时空旅者。大楼的中心是门厅，这里有一间设有外台的咖啡厅。通过门厅，地下的阶梯地势也能接收到来自屋顶花园的光线，并且欣赏到奥胡斯海湾的美丽景色。

展览设计

建筑、自然、文化和历史相互融汇在一起，形成了一种整体性的参观体验。博物馆多年的策展经验及研究将会以一种新方式被发掘出来，用于展示文化历史。莫斯格史前博物馆将以一种让孩子、父母及祖父母们都感到新鲜有趣的方式来培养和促进他们汲取知识。无论他们的出发点是什么，每个人在这里都将有所收获。

通过明亮的天井、阶地和小洞穴般的"房中房"，博物馆能够承办许多主题新颖、类型多样的展览，这些展览结合了科技与作坊式的布局方式，使参观者能够大致了解到考古学家和人类学家是怎样工作的。

建筑材料

所选的建筑材料与整体的建筑表达协调一致；同时，设计还考虑到了声学、经济、技术环境、维护、耐久性、色彩搭配以及可持续性等方面的问题。博物馆的内墙一般是经过喷涂或者保留粗糙表面的混凝土墙。在混凝土大梁之间安装可调节声效的天花板系统，这样也让建筑纵向上的梁保持了可见性。展览室的地面是架高的木材贴面的地板。

建筑的外观主要为巨大的整体混凝土屋顶。屋顶表面覆盖着草坪，草坪中的步行小径同时还具有紧急疏散逃生的功能。

可持续性

建筑和科技之间的互动，是一座建筑要达到美观、舒适、节能这几点的关键之所在。在Henning Larsen建筑师事务所，设计师通过专注于降低建筑能耗这个基本策略，把可持续性的概念落到实处。设计师这样做是相信专注于能耗会在项目的每一个方面都产生全面的品质提升。这

些也都基于一套"用知识去设计"的方法论。项目的基本目标是在三个可持续发展的方面创造价值,即经济、社会和环境。

在莫斯格史前博物馆整体的建筑技术和结构安排上,可持续性一直都是一个重点因素。南向的屋顶表面(屋顶立面)确保了这一节能建筑的计算基础,使该建筑的能效等级能达到一级。

Moesgaard Museum

The new Moesgaard Museum is uniquely located in the hilly landscape of Skåde. With its sloping roofscape of grass, moss and flowers in bright colors the building will appear a powerful visual landmark perceptible even from the sea.

The rectangular shaped roof plane seems to grow out of the landscape and during summer it will form an area for picnics, barbecues, lectures and traditional Midsummer Day's bonfires. Come winter snowfall, the sloping roof will become transformed into the city's best toboggan run.

The interior of the building is designed like a varied terraced landscape inspired by archaeological excavations gradually unearthing the layers of history and exposing lost cities. The visitor can move through a vivid sequence of exhibitions and scientific experiments – like a traveler in time and space. The heart of the building is the foyer with a café and outdoor service. From the foyer, the terraced underworld opens up to the light from the roof garden and the impressive view of the Aarhus Bay.

Exhibition Design

Architecture, nature, culture and history will fuse together into a total experience, and the museum's many years of exhibition experience and research will be drawn upon in a new approach to the presentation of cultural history. Moesgaard Museum will be able to facilitate their knowledge in a way that is interesting to children, parents, and grandparents. There is something for everyone regardless of their point of departure.

With its light courtyards, terraces and small cave-like "houses in the house", the museum will encourage many new and alternative

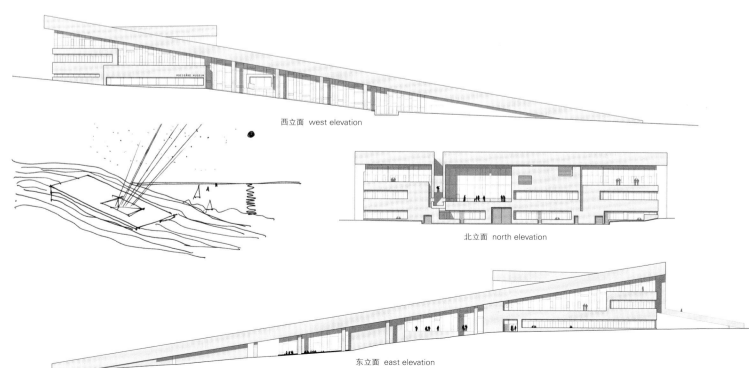

西立面 west elevation

北立面 north elevation

东立面 east elevation

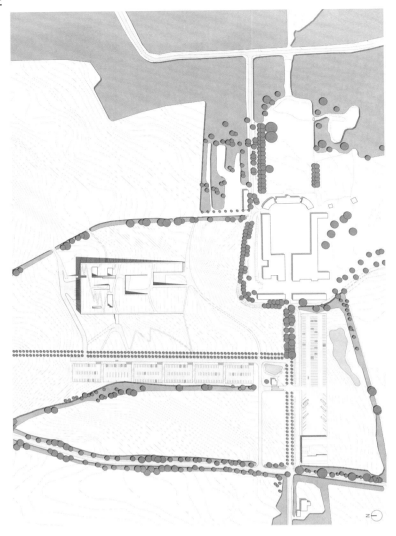

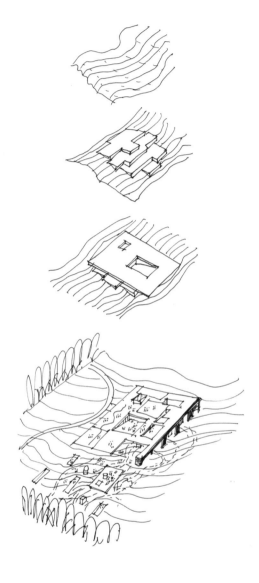

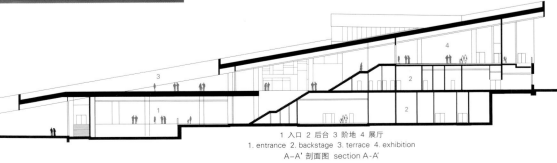

1 入口 2 后台 3 阶地 4 展厅
1. entrance 2. backstage 3. terrace 4. exhibition
A-A' 剖面图 section A-A'

types of exhibitions where the use of technology combined with a more workshop-like arrangement will give the visitors a glimpse of how archaeologists and ethnographers work.

Materials

The materials of the building harmonize with the overall expression of the building and at the same time acoustics, economy, technical settings, maintenance, durability, color options and sustainability are considered. The walls inside of the museum are generally painted or in rough concrete. Acoustically regulating ceilings are mounted between the concrete beams in order for the beams in the longitudinal direction of the building to become visible. The floors of the exhibition rooms are raised floors with a wooden surface.

The exterior of the building is dominated by the large unifying roof surface with a roof ending in concrete. The roof surface itself is covered in grass with paths that live up to the demands of escape routes.

Sustainability

The key to aesthetic, comfortable and energy-efficient buildings is found in the interaction between architecture and technology. At Henning Larsen Architects, we have made the concept of sustainability tangible by focusing on energy reduction as the primary strategy. We have done this with a belief that focusing on energy can create quality all the way round. This is based on the methodology of "Design with Knowledge". The specific gear wheels, agents, have been developed with the objective of creating value for all three aspects of sustainability, that is, economically, socially and environmentally.

Sustainability has been a big factor in the overall architectonic arrangement of Moesgaard Museum. The south facing roof surface (roof facade) ensures the calculated basis for an energy-efficient building, which is able to achieve energy class 1 status.

项目名称：Moesgaard Museum
地点：Aarhus, Denmark
建筑师：Henning Larsen Architects
负责合伙人：Louis Becker
项目经理：Niels Edeltoft
设计主管：Troels Troelsen
建筑师与设计经理：Elizabeth Ø. Balsborg
室内设计团队：Christian Andresen, Karima Andersen, Louise Bay Poulsen, Marie Louise Mangor
项目团队：Birte Bæk, Carsten Fisher, Gitte Edelgren, Greta Lillienau, Hans Vogel, Henrik Vuust, Irma Persson Käll, Johnny Holm Jensen, Julie Daugaard Jensen, Lars Harup, Lars krog Hansen, Magnus Folmer Hansen, Mai Svanholt, Maja Aasted, Martha Lewis, Matthias Lehr, Peter Koch, Sarah Kübler, Stefan Ernst Jensen
工程师：Cowi
景观设计师：Kristine Jensens Tegnestue
承包商：MT Højgaard and Lindpro
甲方：Moesgaard Museum
总建筑面积：16,000m²
设计时间：2005 / 施工时间：2010 / 竣工时间：2014
摄影师：©Jan Kofod Winther (courtesy of the architect) - p.58~59, p.62
©Jens Lindhe (courtesy of the architect) - p.60, p.63, p.64
©Martin Schubert (courtesy of the architect) - p.60~61, p.65, p.66

©Rogvi Johansen(courtesy of the architect)

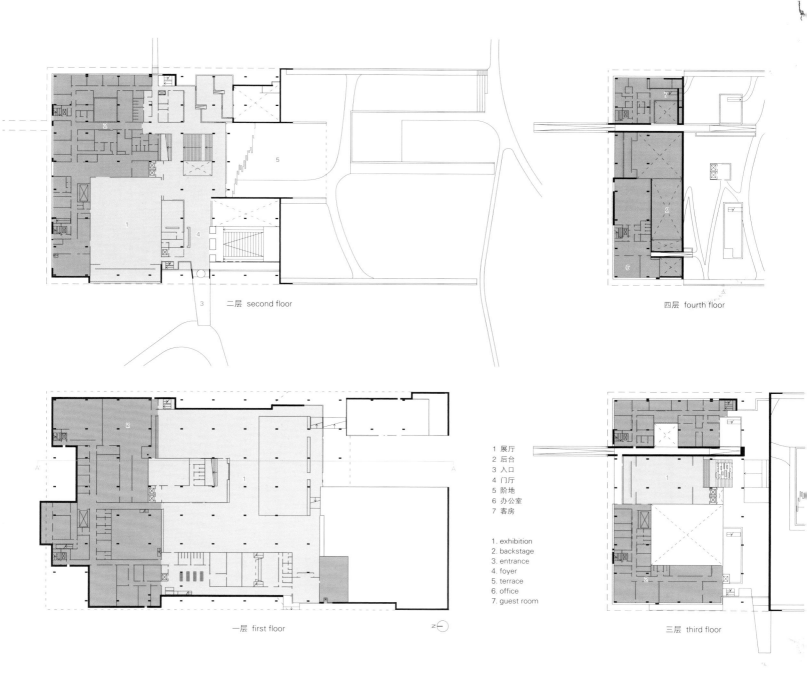

1 展厅
2 后台
3 入口
4 门厅
5 阶地
6 办公室
7 客房

1. exhibition
2. backstage
3. entrance
4. foyer
5. terrace
6. office
7. guest room

二层 second floor

四层 fourth floor

一层 first floor

三层 third floor

景观与建筑 Landscaping and Building

奥比多斯科技园中央大楼
Jorge Mealha

奥比多斯科技园位于里斯本与科英布拉之间的轴心位置，规划目标是将学术研究与商业产品相结合，尤其是创意产业领域。

主楼在设计上定义了一座呈方形环状分布的建筑。其所有公共空间，比如，主会议室/多用途房间、商店、餐厅和微观装配实验室都被有意布置在一层，以增强其内部空间的公共特性。工作区则以模块化网格形式分布于二层，空间使用具有极大的灵活性。这片纯净之地意在表现回廊的典型特征，与中央广场形成了强烈的视觉联系。

这个埋在地下的体量，其立面是穿孔的锈蚀金属表面，让人想起了景观侵蚀的自然过程，限定了该建筑作为一处具有社交性质的公共空间的界限。

从建筑外部是看不到基座的，只能看出一个又长又窄的带状结构，令人想起修道院的围墙，这是典型的当地景观。

建筑使用的三种主要材料是混凝土、钢材和玻璃。基本上，一层主要是粗糙的混凝土表面，用以表现陆地结构。整个一层就像壁炉地面一样，给人以粗糙厚重之感。

与一层相对的是二层，处处运用几何学，讲求精确。一组巨大的金属桁架构成了四处彼此连通的中空空间，建起了一个巨大的悬空回廊。建筑内部的创业启动办公单元以模块的方式分布，占据了这一层的大部分空间。

把一些功能布置在地下具有多重目的。其一是在基地内创造一个绿色表面；其二是降低为建筑内部提供制冷及供暖的暖通空调系统所需的能耗。本项目所使用的大部分建筑材料都是可循环利用的。

Óbidos Technological Park Central Building

On the axis between Lisbon and Coimbra, the Óbidos Technological Park is planned with the aim of linking academic research with business production, especially in the field of creative industries. The project for the main building defines an inhabited topography on which a square shaped ring rests. All the public spaces, such as main meeting/multipurpose room, shops, restaurant and a FAB lab are purposely located on the ground floor in order to reinforce the public character of the interior void. Working areas are distributed on the first floor on a modular grid that allows great flexibility in the use of the space. This piece of pure clarity seeks to reflect the typology of the cloisters, providing a strong visual relationship with the central square.

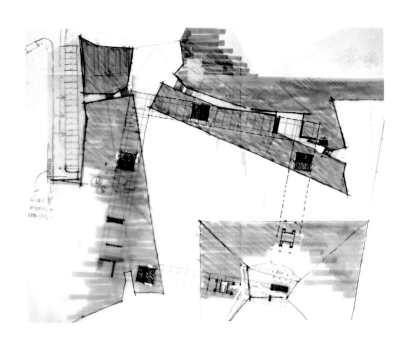
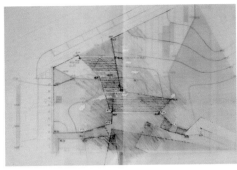
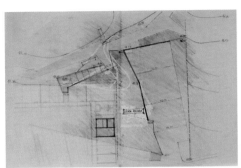

◉	NATIVE EVERGREEN TREES	Ga	TURFSTONE
◉	NATIVE DECIDUOUS TREES	Cg	GRAVEL PATHS
◉	CULTURAL EVERGREEN TREES	Sr	ROLLED PEBBLE
◉	CULTURAL DECIDUOUS TREES	Pb	SCREED PAVEMENT
Pn	NATURAL GRASSLAND	Gr	GRAVEL
Pc	GRASSLAND		BENCH
Ma	HERBACEOUS PLANTS		PAVEMENT LIGHTING

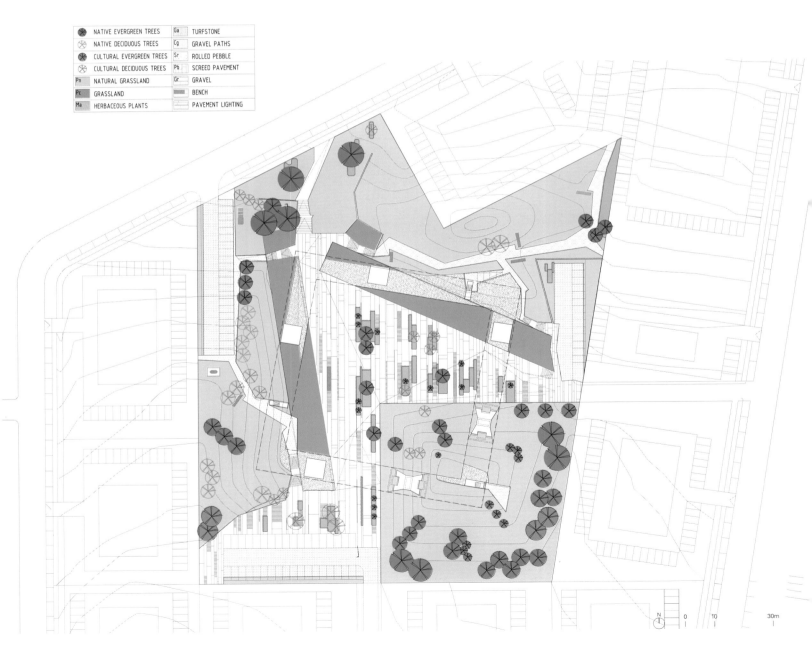

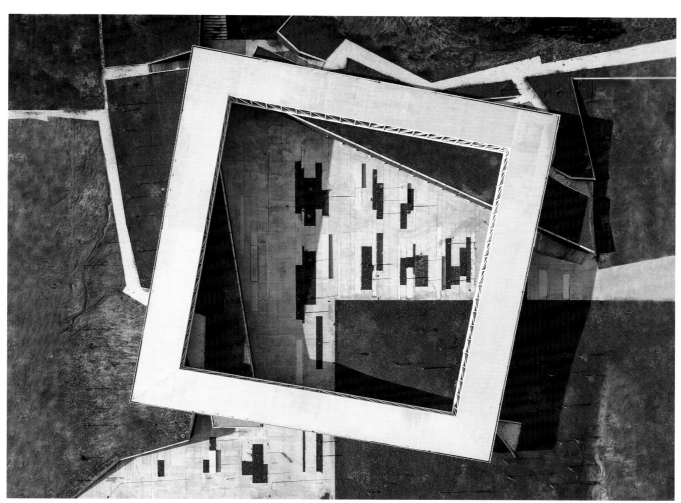

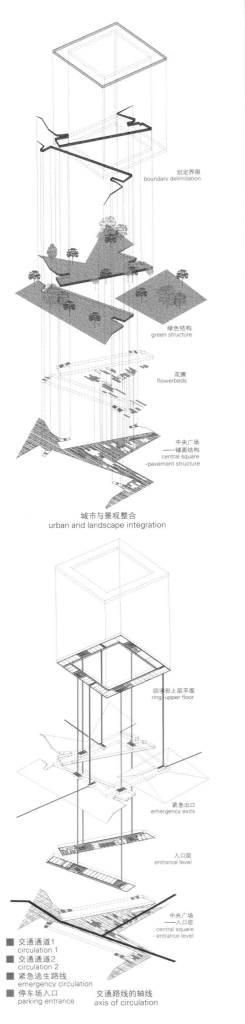

划定界限
boundary delimitation

绿色结构
green structure

花圃
flowerbeds

中央广场
——铺面结构
central square
-pavement structure

城市与景观整合
urban and landscape integration

回字形上层平面
ring upper floor

紧急出口
emergency exits

入口层
entrance level

中央广场
——入口层
central square
- entrance level

■ 交通通道1
　circulation 1
■ 交通通道2
　circulation 2
■ 紧急逃生路线
　emergency circulation
■ 停车场入口
　parking entrance
　　　交通路线的轴线
　　　axis of circulation

The facade of the buried body is formed by a perforated surface of rusty metal that evokes natural processes of landscape erosion and defines the built limits of the convivial public space.

From the outside, the base of the building is hidden and only a long and narrow strip is perceived, evoking the walls of monasteries and convents, which are typical in the landscape of the region. Mainly three materials are used: concrete, steel and glass. Basically, the ground floor is about rough concrete, expressed as a telluric structure.

Opposed to ground floor, the first floor is all about geometry and precision. A set of huge metal trusses, assembled to create four voided and interconnected prisms, builds a large and floating cloister. The structure rims the modularity of the startup office units that occupy most of the space on this floor.

The decision to embed part of the program underneath the landscape aims several goals. One is to increase the green surface within the plot. The other is to decrease energy needs in terms of AVAC systems for cooling or warming the building. A wide range of the materials employed are recyclable.

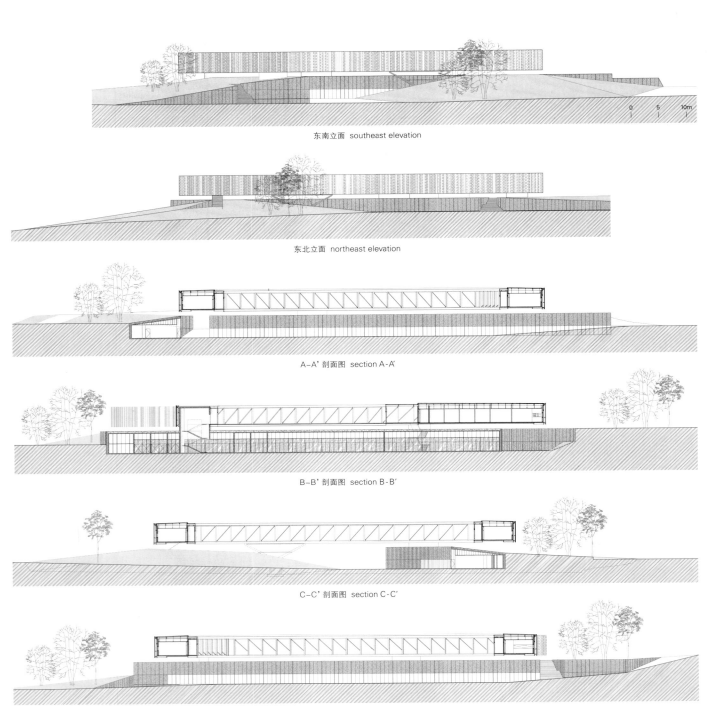

东南立面 southeast elevation

东北立面 northeast elevation

A-A' 剖面图 section A-A'

B-B' 剖面图 section B-B'

C-C' 剖面图 section C-C'

D-D' 剖面图 section D-D'

项目名称：Óbidos Technological Park, Central Building / 地点：Óbidos, Portugal / 建筑师：Jorge Mealha
合作方：建筑师 _ Andreia Baptista, Diogo Oliveira Rosa, Filipa Ferreira da Silva, Gonçalo Silva, Carlos Paulo, Filipa Collot, Inês Novais /
气候、电气、防火顾问 _ Rodrigues Gomes & Associados / 结构工程师 _ José Ferraz & Associados Serviços de Engenharia /
水力学专家 _ S.E. Serviços de Engenharia / 施工单位 _ MRG Engenharia e Construção / 景观设计师 _ Marisa Lavrador
甲方：Parque Tecnológico de Óbidos
用地面积：17,000m² / 总建筑面积：4,096m²
造价：EUR1,076/m² / 设计时间：2011—2012 / 施工时间：2013—2014
摄影师：©João Morgado(courtesy of the architect)

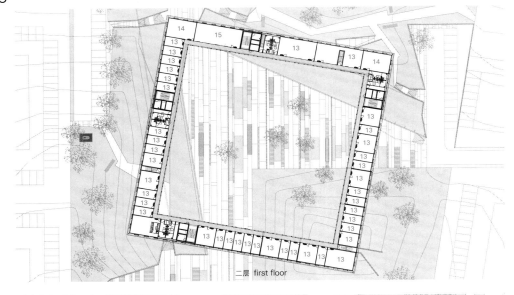

二层 first floor

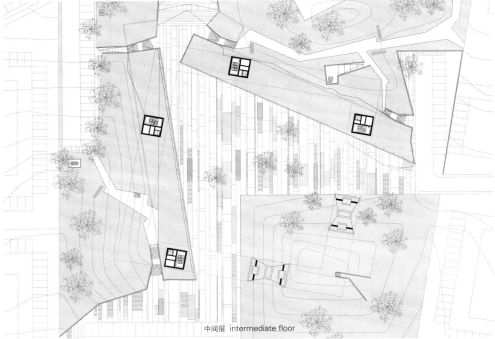

中间层 intermediate floor

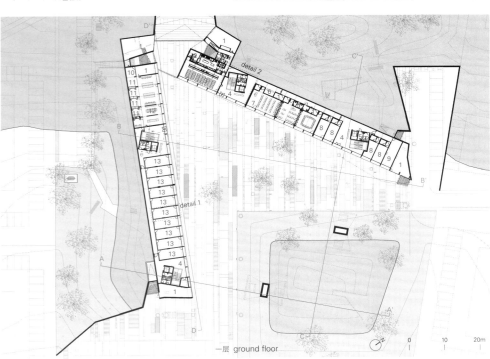

一层 ground floor

1 技术区域	9 数据中心
2 厨房	10 储藏室
3 餐厅	11 办公室
4 中庭	12 微观装配实验室
5 卫生间	13 办公室
6 接待处	14 会议室/培训室
7 会议室/多功能室	15 行政区域
8 商店	16 咖啡休息角

1. technical area	9. data center
2. kitchen	10. storage
3. restaurant	11. office
4. atrium	12. FAB lab
5. restroom	13. office
6. reception	14. meeting & training room
7. meeting & multipurpose room	15. administrative area
8. store	16. coffee break corner

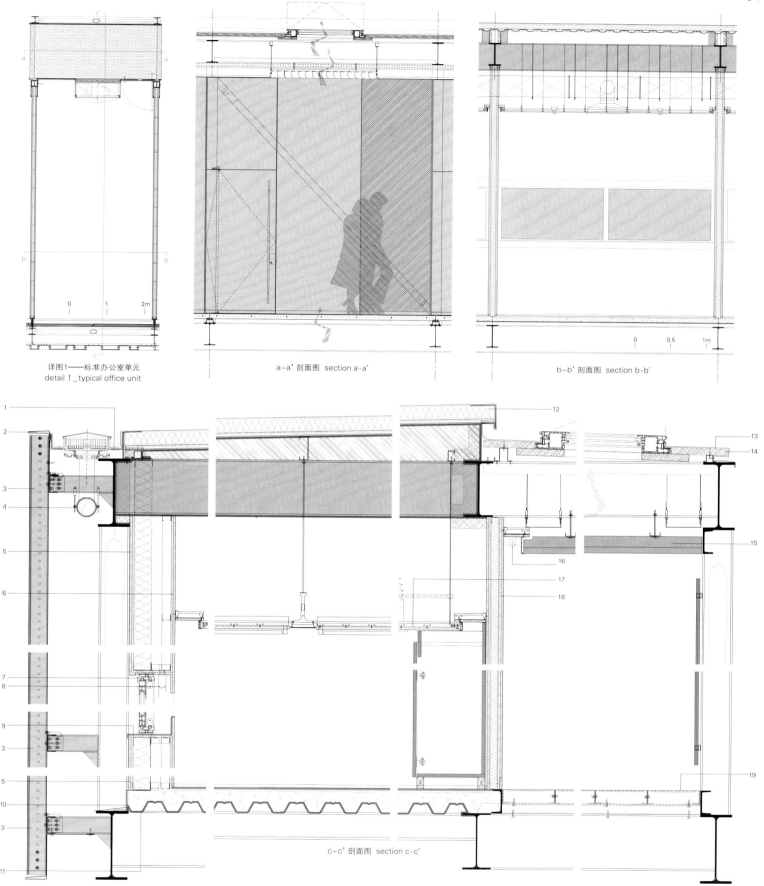

详图1——标准办公室单元 detail 1_typical office unit

a-a' 剖面图 section a-a'

b-b' 剖面图 section b-b'

c-c' 剖面图 section c-c'

1. sheet metal gutter
2. zinc sheet bent to shape, painted white
3. facade plank: meiser sheet metal grating with downward facing perforations and embossing, painted white
4. geberit pluvia roof drainage system
5. exterior wall construction: 15mm permabase cement board, 12.5mm plasterboard, 100mm extruded polystyrene thermal insulation
6. 2×12.5mm plasterboard, painted white
7. double glazing in aluminium frame
8. white roller blind
9. 3mm aluminium sheet bent to shape, painted white
10. 120×80×5mm galvanized steel section welded to main structure
11. thermally insulated composite floor slab
12. roof construction: Sika polymeric waterproof membrane, mineral wool thermal insulation, Sika vapour control layer, 0.70mm trapezoidal profiled sheet
13. Sika plan polymeric waterproof membrane
14. 40mm marine plywood
15. suspended metal ceiling, vertical blade system, painted white
16. florescent light
17. mineral acoustic ceiling with metal substructure
18. wall construction: 2×12.5mm plasterboard, painted black, 40mm thermal insulation, 2×12.5mm plasterboard, painted black
19. meiser sheet metal grating with perforations and embossing in both directions

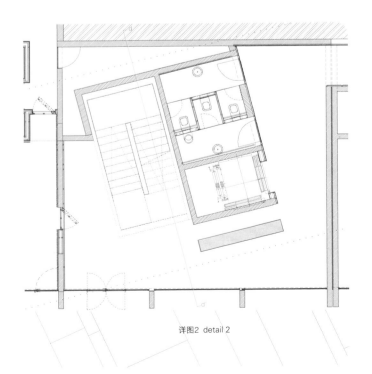

详图2 detail 2

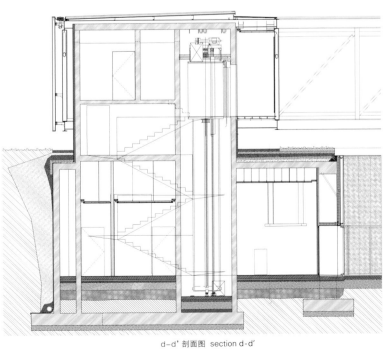

d-d' 剖面图 section d-d'

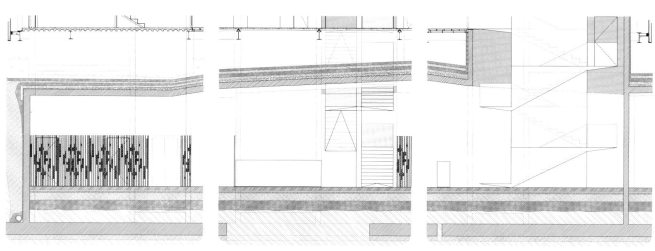

展开立面——入口和楼梯区域（从e到f剖面）
development elevation_entrance and staircase zone(from e to f)

小松科学博物馆

UAO

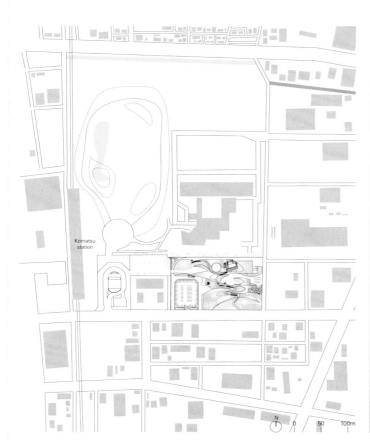

UAO建筑事务所赢得了日本石川小松市的科学博物馆和通信枢纽综合体的设计竞赛，这个综合体坐落在施工和采矿设备的龙头企业——小松制作所的工厂旧址上。

小松市一直以来都是发展生产制造业的工业基地。该市致力于启发"传承制造业灵魂"和"发展儿童在科学领域的兴趣"的思想，并与北陆新干线（一种日本子弹头列车）相结合，进一步成为与其他地区沟通联系的基地。

这个综合体本身是由低矮的四个波浪形结构组成，周围是一些相对低矮的建筑，远处是雄伟的山峰，综合体融入了这样的一个环境中。

科学博物馆在波浪结构的下方，由一个3D圆顶影院、一个科学体验学习中心、一个当地产业升级中心和一个创业中心构成。波浪结构的上表面是屋顶花园，可提高保温效果，并尝试使建筑与景观结合在一起，使这里成为一个可以供人们散步的公共屋顶公园。

人们可以在宽敞的波浪结构的内外随意闲逛，并从多个角度观赏展览。曲面屋顶作为一个蜿蜒的弧形天篷，可控制采光，也可作为排水系统，直接将雨水排到蓄水池用于种植灌溉。可测试风向的花园LED灯遍布整个场地，人们可看出当前的风向。整座建筑有意与科学技术和谐相融，并鼓励参观者在实践中探索更多的科学现象。

该建筑的核心设计理念是波浪形结构的立体扭转和有机的结构动态，使人想起CT扫描（计算机断层扫描）得到的无数横剖面。另外，在地面上弯曲延伸，之后形成波浪形结构和公园的混凝土路面为设计增加了自然、灵敏的元素。

Science Hills Komatsu

We won the competition of Komatsu city, Ishikawa to design a complex of science museum and communication center, which was constructed at the former factory site of a leading company of construction and mining equipment, Komatsu Ltd.

Komatsu city has always developed as an industrial base focusing on manufacturing. It has been requested to enlighten "inheritance of manufacturing spirit" and "children's interest in science", and further take a role as communication's bases with other areas in association with Hokuriku Shinkansen (a Japanese Bullet Train). The complex itself is constructed of four low-rise waves blend-

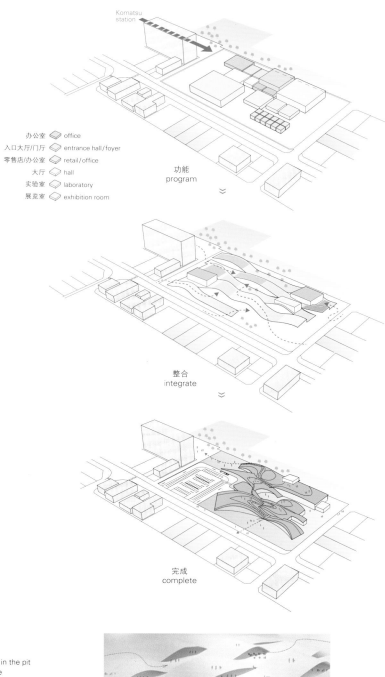

办公室	office
入口大厅/门厅	entrance hall/foyer
零售店/办公室	retail/office
大厅	hall
实验室	laboratory
展览室	exhibition room

功能 program

整合 integrate

完成 complete

1. mesh fence
2. planting
   sprinkling facilities
   lightweight soil
   drainage panel
   drainage sheet
   protective concrete layer
   styrofoam insulation
   wet process tarpaulin
3. air conditioning in the pit
4. ironing concrete
   osmotic concrete reinforcement
   concrete slab
5. blowing non-freon insulation
6. roof illumination
7. air conditioning balloon
8. mortar painting
9. polystyrene form

景观，作为从几何建筑体演化过来的有机建筑
建筑的外形设计为波浪形，其设计目的通过地面与屋顶花园的无缝连接来实现。

Landscape as organic architecture from geometric building
The form of the building was designed in the shape of a wave. The purpose was arranged by the ground being seamlessly connected to a roof garden.

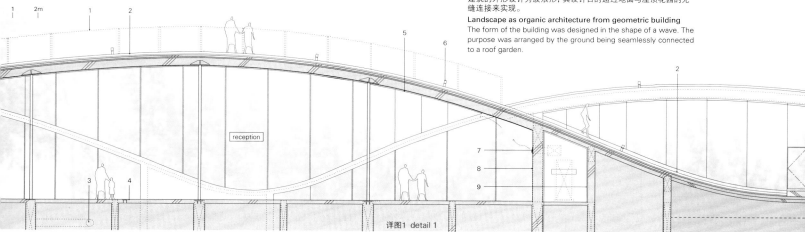

详图1 detail 1

项目名称：Science Hills Komatsu / 地点：Komatsu, Ishikawa, Japan
建筑师：Mario Ito
结构工程师：Kanebako Structural Engineers
施工：Kumagai Gumi + Kaetu Construction Joint Venture
用途：museum
用地面积：14,428.84m² / 建筑面积：6,153.21m² / 有效楼层面积：6,063.03m²
结构：reinforced concrete
外部饰面：exposed concrete, water repellent paint
内部饰面：地面_mortar with metal trowel, surface reinforcing paint /
墙_plasterboard t=12.5, emulsion paint / 顶棚_sprayed sound absorbing urethane foam
设计时间：2011.12—2012.8 / 竣工时间：2013.10
摄影师：©Daici Ano(courtesy of the architect)

结构方案发展模型
structural scheme development modeling

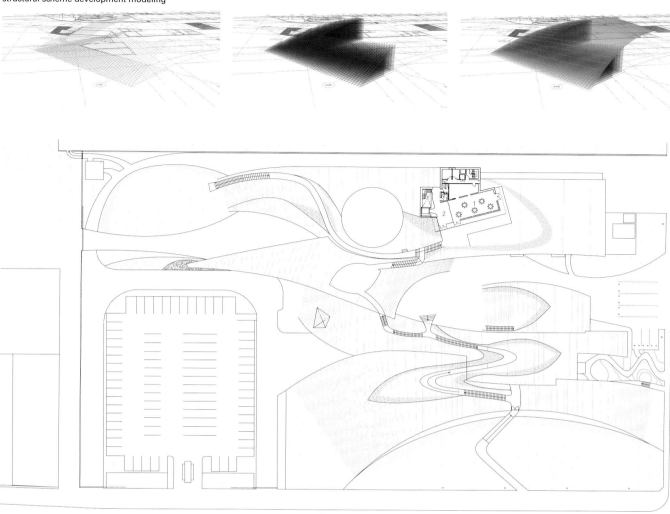

1 咖啡馆&餐厅 2 走廊　1. cafe & restaurant 2. gallery
二层 second floor

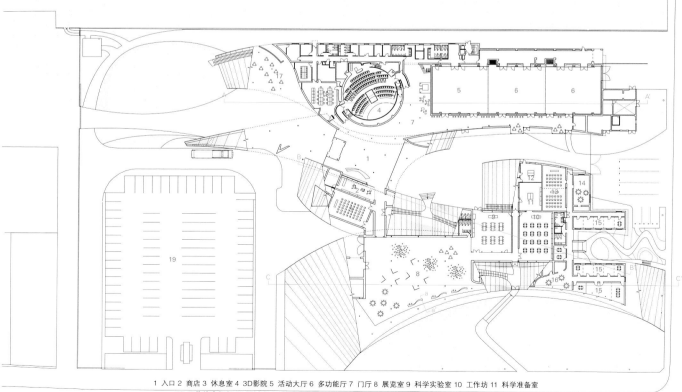

1 入口 2 商店 3 休息室 4 3D影院 5 活动大厅 6 多功能厅 7 门厅 8 展览室 9 科学实验室 10 工作坊 11 科学准备室
12 工作坊准备室 13 研讨室 14 协调员办公室 15 创业室 16 休息厅 17 屋顶庭院 18 办公室 19 停车场
1. entrance 2. shop 3. resting room 4. 3D theater 5. event hall 6. multi-purpose hall 7. lobby 8. exhibiton 9. science lab 10. workshop room 11. science preparation room
12. workshop preparation room 13. seminar room 14. coordinator office 15. incubation room 16. lounge 17. roof courtyard 18. office 19. parking
一层 first floor

ing into the surrounding relatively low-rise buildings, and also into the backdrop of faraway grand peaks. The Science Museum is located under the waves and consists of a 3D dome theater, a science experience learning center, a local industrial promotion center, and an incubation center. There is rooftop gardening at the upper surface of the waves for enhancing insulation efficiency, and it is the attempt to integrate architecture and landscape as a public roof park where people can walk around.

People can stroll freely inside and outside of the broad waves and view exhibits from many aspects. The curve rooftop serves as sweeping canopy controlling light, and also as drains directing rain water into a reservoir for planting irrigation. Wind-detecting LED garden lights are located throughout the whole site and visualize the wind. The entire building is intending to be in harmony with science and to encourage the visitor's various scientific discoveries in practice.

The core design concept has the 3D twist of waves and the organic dynamism which remind limitless number of cross sections of CT scan. Also, the slab sweeping from the ground and then forming waves and park adds a natural and sensitive aspect to the design. UAO

南立面 south elevation

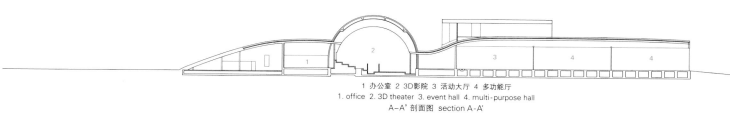

1 办公室 2 3D影院 3 活动大厅 4 多功能厅
1. office 2. 3D theater 3. event hall 4. multi-purpose hall
A-A' 剖面图 section A-A'

1 商店 2 科学实验室 3 工作坊 4 创业室
1. shop 2. science lab 3. workshop room 4. incubation room
B-B' 剖面图 section B-B'

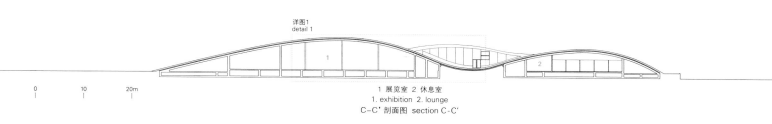

1 展览室 2 休息室
1. exhibition 2. lounge
C-C' 剖面图 section C-C'

1. planting
   sprinkling facilities
   lightweight soil
   drainage panel
   drainage sheet
   protective concrete layer
   styrofoam insulation
   wet process tarpaulin
2. ball void
3. blowing non-freon insulation
4. steel curtain wall
5. mullion steel 38×150
6. air conditioning slit
7. skeleton chamber
8. protective concrete layer
   styrofoam insulation
   wet process tarpaulin
9. float grass
10. liquid-applied membrane waterproofing
11. ironing concrete
    coating for axial sliding
    concrete slab

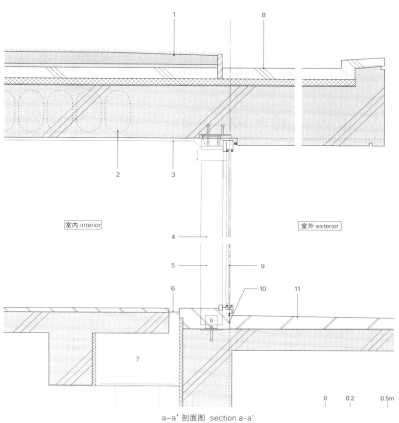

a-a' 剖面图 section a-a'

Stand on the Climate

立足于气候

沙漠中的黑色住宅_Black Desert House/Oller & Pejic Architecture
沙漠庭院住宅_Desert Courtyard House/Wendell Burnette Architects
哈达威住宅_Hadaway House/Patkau Architects
四季住宅_Seasonless House/Casos de Casas
居所与气候_Dwellings and Climate/Aldo Vanini

生态平衡的自然倾向似乎使人们越来越相信，人类对于事物现存状态的任一变更都可能会对其生存带来威胁。但是，历史的全过程就是一个持续适应环境条件的过程。为了适应自然，人类迸发出了对于知识和技术能力的渴望。在这些适应过程中，居所和服饰方面的变化标志着人类学结构和社会关系网方面的转折。

总的来说，尽管有些共同的见解，但人造的城市环境是人类主体给予生存挑战的回应，而这一生存挑战正是由自然环境和气候所带来的。事实上，仅仅是大都市快速增长的步伐，以及日益增长的对于流动性和生产的需求，已经使得城市环境变得相对恶劣。但是，正是在这些自然保留了其原始特点的地方，人类更应该处理已经发生的严峻状况。

除了被人类选作栖息地的传统地理位置以外，在这里所列举的例子告诉我们恶劣的环境条件在很大程度上能够促使人们构思出既兼顾利益又具备良好建筑质量的方案。沙漠和高山环境尤其显现了人类应对极端自然环境的强大能力。

The natural tendency toward homeostasis tends to convince humankind that every modification of the known state of things is to be regarded as a serious threat to its very existence. However, the entire course of history is a long series of continuous adaptations to environmental conditions, adaptations that have contributed to the human desire for knowledge and our technological capacity. Among these adaptations, those in dwellings and clothing mark significant turning points in the anthropological structure and system of social relations.

Common opinion notwithstanding, the artificial urban environment, in general terms, has been the primary human answer to the survival challenges posed by the natural environment and climate. In fact, only the great pace of metropolitan growth and increasing needs for mobility and production continue to make the urban environment relatively hostile. However, it is in places where Nature retains all of its original characteristics that humanity must continue to deal with its severity and rigor.

Far from the traditional geographical sites carefully selected by humanity for habitat-building, the examples proposed herein show the great extent to which critical situations can inspire solutions of extreme interest and architectural quality. Desert and mountain environments in particular represent stages for spectacular demonstrations of the human capability to challenge the most extreme climatic conditions.

居所与气候

居所和服饰，自有人类起，便是贯穿于变革的两大纽带，与人类和气候紧密相连。针对这些地方的气候所衍生出来的解决方案，已经超越了功能的需求，并且开辟了它们自身对于艺术追求的道路。事实上，人类在自然面前应该如何作为，曾被意大利诗人和哲学家贾科莫·莱奥帕尔迪用大篇幅的内容描述过，这些都在他的长诗《金雀花》中有所体现。

芳香的金雀花，
与沙漠为伴，
盛开在维苏威火山的荒芜的山坡上，
它是可怕的高山，万物的毁灭者，
没有花朵装点色彩，没有树木衬托生机，
你撒播着寂寞，
曾经的荒芜被你装扮得生机勃勃，
城市的周围遍布你的身影，
它们曾经是世界的主人，
现在却只能对着坟墓叹息，
它们忍受着寂寞、孤独，
怀念着逝去时代来来往往的路人。[1]

目前看来，人与自然应该尽可能地和谐相处。但是这种友谊关系的频繁出现，就像是恶霸和受害者之间牵强的协议，并且这一现象是由内疚和恐惧引起的。对于人类的未来而言，气候问题是关键问题。对于自身感同身受的经历，人们言过其实的害怕使得他们渐渐忘却了自己强大的适应能力，而这一能力曾经在数千年的气候变化中被验证过。但是，不是所有的人都选择用害怕来面对。颇受争议的物理学家弗里曼·戴森坚信气候变化是一次机遇，可以促进新的、积极的进化。如果有必要，有可能会发生激进的、戏剧性的转变。

以人为本的伦理观念认为，人类是自然界必不可少的一部分。通过人类的思想，生物圈已经获取了掌控自身进化的能力，现在是我们在掌控。人类有权利和义务重建自然，只有这样，人类和生物圈才能共同存活，并且繁荣发展。

对于人类学家来说，最高的价值就是实现人与自然的和谐共存。[2]

随着时间的发展以及气候所带来的危机、不安，人们逐渐形成了一

Dwellings and Climate

Dwellings and clothing, two of the most fundamental nexuses of revolution since the beginnings of humankind, are connected to the relationship between humanity and the climate. The survival solutions for every weather condition produced in those areas have transcended functional needs, opening the way for disciplined pursuits that for their semantic content are what we call art. Actually, the human position before Nature was beautifully and dramatically expressed by Italian poet and philosopher Giacomo Leopardi in his canto *"The Wild Broom or the Flower of the Desert"*:
Fragrant broom,
content with deserts:
here on the arid slope of Vesuvius,
that formidable mountain, the destroyer,
that no other tree or flower adorns,
you scatter your lonely / bushes all around. I've seen before
how you beautify empty places
with your stems, circling the City
once the mistress of the world,
and it seems that with their grave,
silent, aspect they bear witness,
reminding the passer-by
of that lost empire.[1]

Current opinion holds that the human–climate relationship should be as friendly as possible, but the friendship suggested too often appears to resemble the strained accord of bully and victim, occasioned by guilt and fear. Climate concerns are certainly a central issue to the future of mankind, but the exaggerated fear with which they are often experienced seems to induce forgetfulness of our formidable capacity for adaptation, proven in the most extreme climatic variations over thousands of years. However, not everyone has chosen to respond with fear. Controversial physicist Freeman Dyson is strongly convinced that climatic variations can be considered opportunities for new and positive evolution also, if necessary, by radical and dramatic transformations.

The humanist ethic begins with the belief that humans are an essential part of nature. Through human minds the biosphere has acquired the capacity to steer its own evolution, and now we are in charge. Humans have the right and the duty to reconstruct nature so that humans and the biosphere can both survive and prosper.

For humanists, the highest value is harmonious coexistence between humans and nature.[2]

With time and the overcoming of the discomforts and of dangers

沙漠中的黑色住宅，美国加利福尼亚
Black Desert House in California, United States

种观念，那就是将自然拟人化，把它视为一个虚弱的病人，一个友好的、和蔼的存在，要保护它远离人类不负责任的行为。

语言刻薄的阿根廷作家马丁·卡帕罗斯写道：

生态更倾向于指的是黄金时代、古代神话：在那些时代，自然可以尽情地发展自己，不受人类邪恶力量的干扰。当时有高尚的野蛮人，最重要的是，当时是一片荒原，而且没有被社会所破坏。现在我们站在与几千年前同样的场景中，但方式却不一样了：自然被文化所击溃，失乐园被人类的野心所毁灭。还有被锁住的普罗米修斯、通天塔和它的倒塌：没有宗教会选择不惩罚科技的勃勃野心。在享受胜利的原始人犯下了想要知道胜利果实是什么并且吃掉它的愚蠢错误时，他们打破了永恒不变的守则。上帝的守则很明确：如果你能接受我已经制定好规则的自然，那么一切都会非常完美，当你想要施加自己的影响时，你毁了一切。³

不可否认，在许多的事例中，人类的行为已经引发了严重的环境问题。同样的，这些生态失衡现象的受害者也是人类自身，而不是泰然自若的星球。对于此，最适合的答案莫过于放弃对于自然的改造，同时人类也应该意识到科学和技术能够做些什么。

事实上，现如今，建筑的基本元素不是别的，正是人类面对气候挑战所拥有的顽强意志和智慧。毫不夸张地说，建筑物的整体构造系统正是人类用技术给予气候条件的回应。正是由于这个原因，而不是基于当前行星变化所引起的反应，考虑做出变化对我们来说更为合适。看起来非常危险的情况，也有可能是进化和发展的机遇。

此外，抛开人类对于气候的干预，气候自身的循环性质也是为人熟知。古代的气候事件如古新世－始新世极热事件，始新世极热事件，或者是距离相对较近的中世纪暖期，都发生在人类进行技术活动之前。正如戴森在他四部《异端邪说》中的一部里提到的那样，生物圈在过去也是不断地发生着变化，在将来它还会持续改变。我们想让它停止改变的想法是一种具有危险性的错觉。想要弄清楚进化、非保守性以及必然变更就需要用发展来验证，不仅是在宏观世界中，同时也反映在建筑设计中，反映在更先进的环境解决方案中。

我们已经意识到了气候变化的规模，这次的气候变化是毋庸置疑的，也是在意料之中的，它影响了整个星球。要想改变这种现状就需要实现人类学上的变更。我们需要重新思考基本生活质量观念、资源分配与

to which the climate exposed them daily, humans have developed an anthropomorphic concept of nature as a sort of enfeebled ward, a fragile, gentle entity to be protected from human irresponsibility.

The caustic Argentinean writer Martín Caparrós writes,

The ecology likes to refer to a golden age, to an ancient myth: happy times in which nature could develop itself without the interference of human evil. There were noble savages, but above all there was a good wilderness: and this had not yet been corrupted by society. It's the same scene we stand once again, in many different ways, from thousands of years ago: nature defeated by culture, Lost Paradise, destroyed by human ambitions. Chained Prometheus, Babel and its collapse: there is no religion that does not punish the ambition of the technique. When the original man who lives in the most glorious triumph of nature commits the folly of wanting to know and eat the fruit, he breaks the natural/divine order made for eternity, against the change. The order of God was very clear: everything will be perfect if you accept the nature that I have done to give my rules; you ruin everything, from the moment you try to impose your own. ³

It is undeniable that in too many cases the actions of humanity have created severe environmental situations, but it is equally true that the main victim of these imbalances has been humankind itself, rather than the imperturbable planet, and that the most suitable answer cannot be to give up transformation of the environment, but to do so with full awareness of what science and technology are capable of.

In fact, what today may appear to be elemental archetypes of architecture are nothing but answers produced by the stubborn will and ingenuity of humanity to challenges posed by climate. It is no exaggeration to say that the entire tectonic system of architecture was established as a technological response to climatic conditions. For this reason, rather than react in a conservative manner to current planetary change, it would be more appropriate to consider change, even in its most seemingly threatening aspects, as an opportunity for evolution and development.

Moreover, human climate intervention aside, the cyclical nature of the planet's climate is well known. Ancient climate events such as the PETM (Paleocene–Eocene Thermal Maximum), ETM1 (Eocene Thermal Maximum 1), or the more recent MWP (Medieval Warming Period), occurred well before the remarkable technological activity of man. As Dyson stated in one of his four heresies, *"the biosphere was constantly changing in the past, and it will be constantly changing in the future. The idea that we can put a stop to change is a dangerous illusion".* Thus, the need to figure out an evolutionary

利用以及科学与技术研究的目标。我们在这里所列举的工程实例,以一个浓缩的视角向我们证明了,品质、智慧以及优雅是人与自然和谐相处的关键。

尽管莱奥帕尔迪的悲观情绪被回避掉了,但这些住所在一定程度上让人回忆起了诗歌中金雀花生长的环境。在自然中孤独生长,陌生而又存在着潜在的敌意,承受着自然所带来的冷酷——无论是位于海岸线还是沙漠之中,它们自身与周围环境都形成了鲜明的对比。

有的观点认为此处的建筑代表了人与自然共存的一种特殊情况,它是以特殊的隔离环境和人类富有的资源为前提的,这一点是颇具争议的。这一建议只是对于众多与自然还不和谐地区的肤浅认识。

下面一些工程,尽管被认为处于未被污染的地理位置,但它们拒绝过分简单化的模拟,并且宣称它们的义务就是要从地质构造以及人类学方面处理与自然的关系。在当前的形势和技术条件下,它们代表了一种传统的处理气候变化的方式,通过承认其严酷性,并采取能够维持与自然可持续关系的一切措施保护自己。

要想处理好与自然的关系,不仅要从行为上,也要注重精神和心理上的联系。奥勒·佩伊奇建筑师事务所在加利福尼亚约书亚树国家公园附近建造的沙漠中的黑色住宅,表面主要是黑色的,并且使用了玻璃幕墙。这主要是应客户的要求,客户要求把房子设计得像庇护所一样:一个能够躲避刺眼阳光的庇护所,一个在沙漠的夜晚能消失、与黑暗融为一体的庇护所。房子建立在一块平坦的山地上,周围围绕着一些岩石,有些岩石还穿透了房屋。

跟沙漠中的黑色住宅很相似的是由温德尔·伯内特建筑师事务所设计的沙漠庭院住宅。这一建筑与金雀花在恶劣环境中的濒危情况遥相呼应,它位于亚利桑那州的索诺拉沙漠。建筑师通过材料之间的结合以及使用非正交的几何图形,使得建筑与环境紧密结合,并形成了强烈的明暗对比。如同沙漠中的黑色住宅一样,建筑师并没有将房子构思成一个密闭的场所,而是设计成了路径系统,以保持与自然的连通性。

一个更加与众不同、更具创意性和勇气的设计当属位于西班牙卡斯特利翁省由卡萨斯建筑师事务所设计的四季住宅。与之前的例子有所不同,该建筑并不是建立在相对僻静的环境中,而是建立在繁华的都市之中,以反映出不同的生活方式。正如它的名字所表明的那样,四季

and non-conservative relationship with these inevitable transformations requires the development, not only at the macro scale, but also at the more limited scale of architectural design, of more advanced environmental control solutions.

Although we are well aware that the size of the undeniable, but not surprising, climatic changes that affect the entire planet requires an anthropological transformation in which we reconsider fundamental concepts of quality of life, the use and distribution of resources, and the aims of science and technological research, the examples shown here suggest on a smaller scale that quality, intelligence and elegance are the keys to a solution for a good relationship with nature and the climate.

Although they shun the cosmic pessimism of Leopardi's vision, these houses nevertheless recall in some way the poetic situation of the wild broom, isolated in nature that is alien and potentially hostile, given the hardness of the climate – whether ocean coastline or the desert – but able to establish a fascinating contrast between itself and the context.

It can be argued that the buildings here examined represent specific situations in which coexistence between the natural and the human is guaranteed by privileged isolation and the wealth of the dweller, but such a suggestion would be a superficial assessment considering countless similar situations that have not found a similar harmony with the context.

Some of the following projects, although located in geographical sites that can be considered practically uncontaminated, reject a simplistic mimetic approach, proudly claiming the right/duty to deal with nature in an unequivocally tectonic and anthropogenic vocabulary. They represent, in contemporary form and technology, a traditional way of dealing with the climate, defending themselves by acknowledging its harshness and implementing all measures required for a sustainable relation with the natural environment.

Dealing with the environment implies not just physical, but also mental and psychological relation. The choice of Oller & Pejic Architecture to design the Black Desert House as an assembly of plain black surfaces and glass pans near Joshua Tree National Park in California was dictated by the request of the client "to build the house like a shadow": a shadow as a virtual shelter from the harsh sunlight, a shadow that would disappear, dematerialising at night in the deep darkness of the desert. The house rests on a mesa, part of the rocks that surround it, penetrable like them.

Similarly to the Black Desert House, the Desert Courtyard House by Wendell Burnette Architects echoes the theatrical situation of

哈达威住宅,加拿大不列颠哥伦比亚
Hadaway House in British Columbia, Canada

住宅的设计目的是替那些"更在乎环境而非房子"的人们营造出家庭的、个人的气候,建造出与全球居民栖息环境不断连通的居所,并且减少在此地区复制气候的可能性,因为这往往会引起冲突。房子仍处于施工之中,希望能够建造成让住户可以享受更高层次能源效率的居所。换句话说,这种树脂材料和玻璃材料制成的半透明外壳使得建筑物呈现出一种"清澈"的状态,是一种能够较好适应环境、气候变化以及人类需要的范例。

对于人类居所来说,另一传统的巨大挑战就是四季的气候变化。尤其是高山环境会使人体验到差异巨大的气温变化,使人苦恼于冬天冰雪积压给房屋造成的巨大负荷。哈达威住宅位于不列颠哥伦比亚的山上,帕特考建筑师事务所利用一个倾斜角度很大的屋顶完善了山体原型的解决方案,并将这种几何结构应用到了建筑的所有表面上。结果就是它演变成了人们生活的机器,木板等保温材料的使用使得房屋与寒冷的环境进一步隔绝,大量混凝土墙的使用也使得房屋内的温度进一步得到保证。

以上所有的例子都彰显了人类后天形成的应付极端环境的能力,也表明了正确利用适宜的技术可以得出最小程度上致气候改变的方案。也正是因为有这些努力,曾经被视为是任务的建筑设计现在已成为对特殊生活方式的追求,希望能够营造出一种非都市自然环境。

the wild broom on the verge of hostile nature, here represented by Arizona's Sonora Desert. The building is merged into the place through the incorporation of materials taken from the same site and through the use of non-orthogonal geometries that produce a strong contrast between light and shadow. As in the Black Desert House, the plan of the house is not conceived as a closed box, but as a system of paths in continuity with the natural site.

A different, more innovative and courageous approach distinguishes the Seasonless House, in Vinaròs, Castelló, Spain, by Casos de Casas Architects. Unlike the previous examples, the structure must deal not with an isolated site, but a peculiar urban context planned to host different ways of life. As its name indicates, the Seasonless House is designed to create its own personal, domestic climate for people who inhabit "its practices and not the house", a device in continuity with the global habits of its inhabitant and the possibility of reproducing other climates that can contradict the ones that already exist in Vinaròs. The house is a project in progress, built around a "waiting structure" to be developed "for when the client attains a greater level of energy efficiency". In other words, this translucent shell of polycarbonate and glass represents a liquid approach to architecture, an exemplification of the concept of adaptation to a continuously changing environment, climate, and set of inhabitant needs.

Another traditional great challenge for human dwellings has been wide seasonal climatic variations. In particular, the mountain environment exposes the dweller to a vexing range of temperatures and to snow-induced structural stress during the winter. Hadaway House is located in the mountains of British Columbia. Patkau Architects exacerbate the archetypal mountain solution of a strongly sloped roof, applying this geometry to all the building's surfaces. The result is a kind of radical machine for living whose relationship with the climatic excursion is further facilitated by insulating materials such as wood planks and temperature dampening through massive concrete walls.

All the above examples demonstrate the acquired human capacity to cope with the most rigid contextual conditions and show that a proper use of technology can render less scary those climate change scenarios that are often represented as apocalyptic. Thanks to such efforts, what was once endured as hard necessity has now become a privileged lifestyle choice linked to a desire to deal with a non-urban environment. Aldo Vanini

1. Leopardi, G., *Canti*, Firenze, 1845, Translation by A.S. Kline, 2003.
2. Dyson, F., *Many Colored Glass: Reflections on the Place of Life in the Universe*, University of Virginia Press, 2007.
3. Caparrós, M., *Contra el cambio*, Anagrama, Barcelona, 2010.

沙漠中的黑色住宅
Oller & Pejic Architecture

20世纪60年代,当这一地区初次被规划为发展区时,这一场地就被划定了等级。建筑师通过将周围的岩石磨平以及将凹陷处填平使其变得相对平坦。想要解决此地断壁残垣的问题不仅花费高而且也是难以实现的。想要在建设新房子的过程中解决这问题更是一个极大的挑战。房子被建立在悬崖峭壁边,有360°全方位的开阔视野,望向路边的视线被巨石遮挡。

这一漫长的研究过程从客户向建筑师展示其所感兴趣的房屋图片开始。与西南部沙漠内的实际的中世纪现代住宅相比,大部分都是外形与空间语言更具备侵略性的风格。

让我们来回顾一下先例,看之前的建筑师们在相似的地形条件下是如何进行设计的。可以发现这些建筑师们或是将建筑与风景融为一体,比如,弗兰克·劳埃德·赖特以及鲁道夫·辛德勒的作品,或者是像密斯·凡·德·罗的欧洲现代主义一样,让建筑与景观保持一定的距离。在这片原始的土地上,极简主义的方案更受人们追捧,因此本案建筑师决定将美国西部的大地艺术风俗作为建筑设计更好的出发点,并且将两种倾向与这种关系结合结合起来——一方面,让建筑紧贴地面,另一方面,又让它们保持各自的不同之处。

建筑将会替代被毁坏的土地,不过不是以山脉的形式,而是以一片阴影的形式出现或者是看起来像岩石背向太阳的那一面。给人感觉就好像是岩石被移开了,只留下一块闪烁的黑曜石碎片。

在现场想好设计概念后,建筑师开始重新设计空间及住宅内的交通流线。建筑师希望当人们在房间里进行亲身体验时,能够想起人生中的一次户外旅行。从客厅到卧室,所有的房间都被设计成直线形,厨房和餐厅位于中间,所有的这些都围绕一个内部庭院设置,这个庭院成为从入口开始的序列空间的中间空间,同时也是一个免受严酷气候侵袭的室外空间。由马克设计的客厅从总体上来看像一个别致的睡袋,它通过

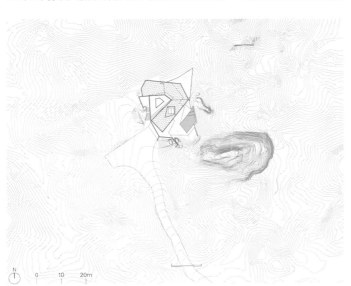

	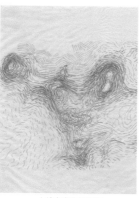	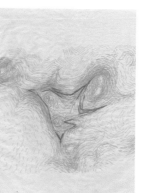	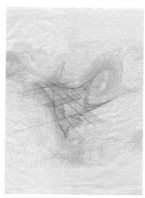	
地形分析 topography analysis	土地密度分析叠加 overlay of landform intensity analysis	不同等级场地人类干预叠加 overlay of human intervention in site grading	辐合线叠加 overlay of convergence lines	与最初的几何形状设计的复合 composite with initial geometry developed

坚实的土墙嵌在了山坡中，人们可以背靠它欣赏远处的地平线，这个空间可以说是一个真正的露营地，在西南地区设在峭壁上的住宅中可以找到建筑先例。

室内暗淡的颜色给人一种原始洞穴的感觉，在白天，当房子的内部变得越来越模糊时，这种感觉就更加明显了。在晚上，房子会消失在夜色之中，同时柔和的灯光和星光在无形中成了沉思的背景。

如果没有整个团队的持续努力，就不会有这个项目的建成。这个项目是马克和米歇尔以及其他建筑师们共同努力的结果。建筑师们的耐心和奉献以及阿维安·罗杰及其分承包商对于项目的成功起到了至关重要的作用。每个为这个项目忙碌的人都知道这是一件不同寻常的事情，都为项目竣工付出了难以想象的努力。

Black Desert House

The site had been graded in the 1960's when the area was first subdivided for development. A small flat pad had been created by flattening several rock outcroppings and filing in a saddle between the outcroppings. To try to reverse this scar would have been cost prohibitive and ultimately impossible. It would be a further challenge to try to address this in the design of the new house. The house would be located on a precipice with almost 360 degree views to the horizon and a large boulder blocking views back to the road.

A long process of research began with the clients showing us images of houses they found intriguing – mostly contemporary houses that showed a more aggressive formal and spatial language than the mid-century modern homes that have become the de-facto style of the desert southwest.

We looked back at precedents for how architects have dealt with houses located in similar topography and found that generally they either sought to integrate the built work into the landscape, as in the work of Frank Lloyd Wright and later Rudolf Shindler or to

项目名称：Black Desert House
地点：Yucca Valley, California, United States
建筑师：Oller & Pejic Architecture
室内设计：Marc Atlan Design, Inc.
工程师：Castillo Engineering
承包商：Moderne Builders
甲方：Marc & Michele Atlan
用地面积：10 235m²
有效楼层面积：451.8m²
设计时间：2009—2010
建造时间：2010—2012
摄影师：©Marc Angeles (courtesy of the architect)

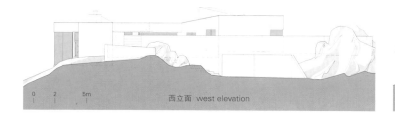

西立面 west elevation

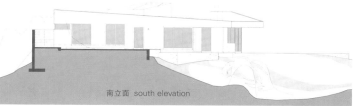

南立面 south elevation

东立面 east elevation

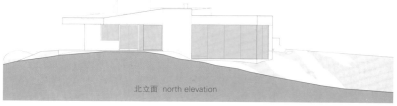

北立面 north elevation

1 车行道	1. driveway
2 泳池	2. pool
3 天井	3. patio
4 庭院	4. courtyard
5 客厅	5. living room
6 厨房	6. kitchen
7 主卧	7. master bedroom
8 大厅	8. hall
9 杂物间	9. utility room
10 卧室	10. bedroom

一层 first floor

1 泳池	1. pool
2 车库	2. carport
3 机械室	3. mechanical

地下一层 first floor below ground

hold the architecture aloof from the landscape as in the European modernist tradition of Mies van der Rohe. While on a completely virgin site, the lightly treading minimalist approach would be preferred, here we decided that the Western American tradition of Land Art would serve as a better starting point, marrying the two tendencies in a tense relationship with the house clawing the ground for purchase while maintaining its otherness.

The house would replace the missing mountain that was scraped away, but not as a mountain, but a shadow or negative of the rock; what was found once the rock was removed, a hard glinting obsidian shard.

Concept in place, we began fleshing out the spaces and movement through the house. We wanted the experience of navigating the house to remind one of traversing the site outside. The rooms are arranged in a linear sequence from living room to bedrooms with the kitchen and dining in the middle, all wrapping around a inner courtyard which adds a crucial intermediate space in the entry sequence and a protected exterior space in the harsh climate.

The living room was summed up succinctly by Marc as a chic sleeping bag. The space, recessed into the hillside with a solid earthen wall to lean your back against as you survey the horizon is a literal campsite which finds its precedent in the native cliff dwellings of the southwest.

The dark color of the house interior adds to the primordial cave-like feeling. During the day, the interior of the house recedes and the views are more pronounced. At night the house completely dematerializes and the muted lighting and stars outside blend to form an infinite backdrop for contemplation.

The project would never have come about without the continued efforts of the entire team. The design was a collaborative effort between Marc and Michele and the architects. The patience and dedication of the builder, Avian Rogers and her subcontractors was crucial to the success of the project. Everyone who worked on the project knew it was something out of the ordinary and put forth incredible effort to see it completed. Oller & Pejic Architecture

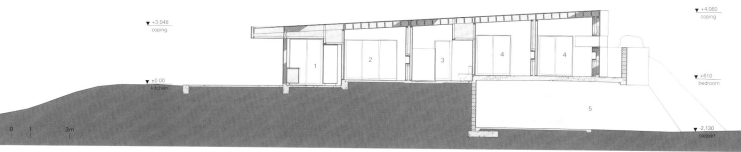

1 厨房 2 主卧 3 浴室 4 卧室 5 车库 1. kitchen 2. master bedroom 3. bathroom 4. bedroom 5. carport
A-A' 剖面图 section A-A'

沙漠庭院住宅
Wendell Burnette Architects

大地和天空

该项目场地被坚硬的花岗岩和高高耸立的仙人掌环绕四周,由于经历沙漠的常年洗刷,这里只剩下一座遍布仙人掌的山脊。到达建筑场地还需要一段长长的车程,蜿蜒向西的广阔远景布局及远方层峦叠嶂的山脉,使得夜晚的这里看上去似乎是在上演一部亚利桑那州日落的话剧。由于建筑高居于生活社区之上,入口大门非常重要——通过这里可以像一道影子一样隐入沙漠之下更深处的设施之中。

当你踏上精致的地板在房屋中穿行,你会有置身禅意花园之感。在该场地的制高点,我们会很高兴看到一座巨大的箭形建筑指向西方、一组特别的火山岩石,还有一株巨大的多枝的树形仙人掌矗立其间。第一次站在这个地方,我们可以立即感受到强烈的保护欲,想要珍存这一珍贵的原始沙漠景观的缩影,还有外面那一片无边无际的广阔天空。

静谧的沙漠庭院景象开始慢慢浮现出来。庭院的设计理念带来的舒适感觉——包括空气、光线、私密性及宁静之感——是所有人对于家的共同渴望,深深地吸引着客户。建筑师开始设想"庭院的院墙能否就地生长出来,而不进行不必要的运进运出"。

设计团队通过土壤样品检测,发现了该项目挖掘过程中被挖掉的泥土可以在建造主结构时当作夯土墙使用,这是一种古老的建筑方法。先设置间隔较大的木模板,在上面铺设91cm厚的土,然后将干燥的泥土和3%~8%水泥的混合物压实成30cm厚的密实的蓄热体,反复堆叠直到达到墙的高度。在地面需要一个稳定的基座,在上部则需要一个顶盖来防止雨水侵蚀。建筑师就在洪水水位处建起了必要的地基,并将其扩大延伸至庭院,作为北面厚重围墙之外的一个主楼层,以最基本的形式观望着广阔的大地和天空——基于广袤大地之上的巨大基座。

基座使用了与墙壁、地面、坡道、阶梯甚至石凳都完全相同的材料现场浇筑而成,会使人感觉到它们都是一块巨大的石头的一部分。佛得角河最终会汇流到索尔特河,在山谷的最低点聚集了世界上最难齐集的砂石和水泥,当地的混凝土生产工业也因此获得了丰厚的收获。建筑师选择了"高速公路混凝土拌合物"以及超大尺寸的3.8cm的集料,并混合了少部分无机颜料——生赭,用于该项目的建造。建筑师想要通过基座表面表现砾石、卵石、碎石、水泥材料的特质,使该基座成为观察该地区地质时代的一个窗口。

整体高度遵循设计准则,因此地面比自然地面高出了7.3m,以几乎觉察不到的变化方式分段螺旋延伸,庭院向西面开放。

Desert Courtyard House

Earth and Sky

The site is a peninsula of granite outcroppings and towering Saguaro cacti surrounded on all sides by deep perennial desert washes except for a single spit of land affording access from an Ocotillo studded ridge above. The building site, further down a long private drive, levels out toward the west into an edge condition dominated by an expansive vista – layers and layers of distant

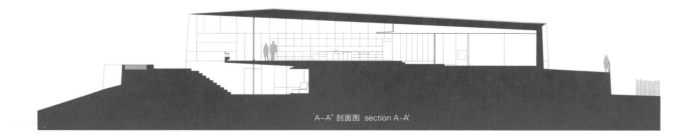
A-A' 剖面图 section A-A'

mountain ranges – that in the evening seems to epitomize the drama of the Arizona Sunset. Due to the elevation of the site beneath the community's gaze and the entry gate at the road it became important to us – to recede the house as a deep shadow – into the depth and complexity of the desert floor below.

When your feet begin to move across this delicate floor you feel as though you have entered a Zen garden. At the highest point of the site we took delight in a large arrow-shaped granite boulder pointed west, a peculiar group of volcanic rocks and a large multi-armed Saguaro between. Standing in this place for the first time, we felt immediately compelled to hold and preserve a microcosm of this precious primordial desert landscape including an equally infinite piece of its indomitable sky.

The form for a serene desert courtyard slowly began to emerge. The courtyard concept intrigued our client as the comforts it offered – air, light, privacy, security, and tranquility – and they were the ones universally desired in a home. We began to ask the question "whether the courtyard walls could literally grow from the

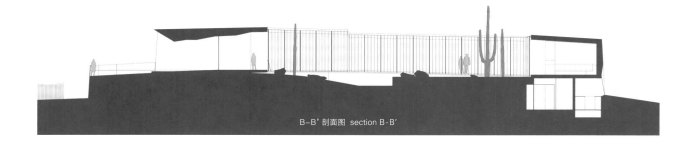

B-B' 剖面图 section B-B'

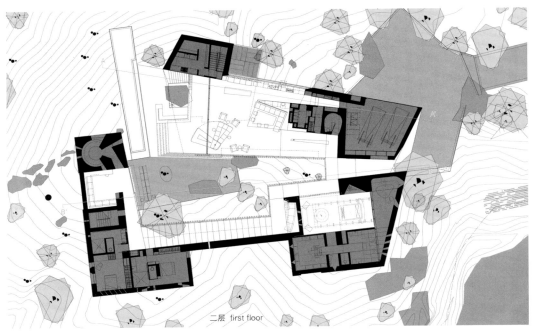

二层 first floor

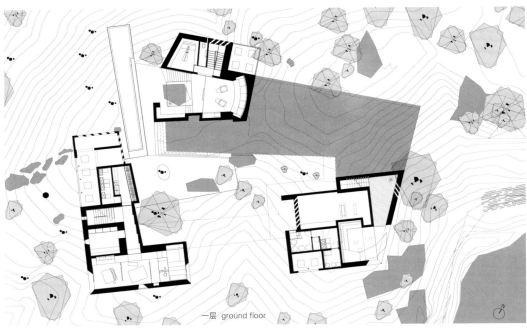

一层 ground floor

site itself, from the site excavation with no import and no export required?"

We discovered through sample testing that the soil from our site was uniquely suited for rammed earth, one of the oldest methods of construction. Wooden molds held wide apart accept 91 cm high layers of earth in lifts, then a dry mix of dirt and cement (3%~8%) is compacted down with pressure pounding into a 30 cm layer of dense thermal mass, lift by lift until the height of the wall is achieved. It requires a stable footing at the ground and a hat for protection from rain and erosion. We raised the requisite foundation just above the flood plane as a base and then expanded it into the courtyard as a piano nobile, beyond the thick perimeter earthen walls, as the most elemental form from which to view the expansive qualities of land and sky – a massive land based land scaled - plinth.

The plinth was cast in place with one material throughout such that a wall, a floor, a ramp, a step, or a bench could be experienced as part of one contiguous stone. The Verde River eventually connects to the Salt River, which collectively tumbles some of the world's hardest aggregate through the lowest point of the valley, where along with sand and cement, it is harvested for locally produced concrete. A "highway concrete mix" with oversized 3.8 cm" aggregate was specifically selected for this project and mixed with a small percentage of the earth pigment – raw umber. We wanted to work the surfaces of the plinth in order to reveal the composite qualities of the material, sand, conglomerate gravel, pebbles, broken stone, in a cement matrix, and consequently a window into the geologic time of this place.

The overall height of the landform follows the design guidelines and therefore the ground at precisely 7.3m above natural grade in a segmented monocline that spirals almost imperceptibly up and around and out where the solid mass of the courtyard form opens up to the distant west.

项目名称：Desert Courtyard House
地点：Scottsdale, Arizona, U.S.
建筑师：Wendell Burnette Architects
设计负责人：Wendell Burnette
项目经理/设计主要合伙人：Thamarit Suchart
项目团队：Jena Rimkus, Matthew G. Trzebiatowski, Scott Roeder, Brianna Tovsen, Chris Flodin, Colin Bruce
机械工程师：Kunka Engineering, Inc.
电气工程师：Associated Engineering
土木工程师：Rick Engineering
照明：Ljusarkitektur P&O AB
景观设计：Debra Dusenberry Landscape Design
室内设计/家具设计：Leavitt – Weaver, Inc
声音/视觉/隔声设计：Wardin Cockriel Associates
承包商：The Construction Zone, Ltd.
用地面积：7200m²
竣工时间：2014
摄影师：©Bill Timmerman (courtesy of the architect)

哈达威住宅
Patkau Architects

这座雪国的房子坐落在可以眺望不列颠哥伦比亚省西南方惠斯勒山谷全景的西北坡上。这个地点是一个问题诸多的楔形坡,坡顶上的道边仅能设置一个车库和一个狭窄入口。

两个主要因素的结合决定了房子的外部造型:第一个是容许的建筑面积和高度;第二个是让雪从屋顶上滑下来存储到场地内合适的位置的需要。这个雕塑般的体量从突出的建筑足迹与折叠状覆雪屋顶的交叉区域中浮现出来,而对内部空间的利用保持了建筑物形状内在的塑性。

主要的楼层是一个大空间,其中容纳了起居室、餐厅、厨房以及一个室外平台,所有这些空间都面向山谷,可以一览山谷美景。一条垂直的裂缝在最高的屋脊下伸展,将这个弯曲的空间一分为二并给最深处的空间带来光明。楼梯设在裂缝中,一座连接桥在上层位置横跨裂缝连接主卧和书房。最下面一层是更私密的空间,包括客房、另一间起居室以及一个大的服务区域。从车库入口可以直接进入房子,这个服务空间提供这里基本的生活所需。在雪国,一般把湿衣服挂起来晾干或者直接扔去洗衣店。滑雪者可以把他们日常活动所需的随身用品储存在这里。另一部楼梯连接着最下面一层和房屋下面的室外庭院,这里是除了入口之外这个陡峭场地内唯一的一处平地。

整体系统和框架系统结合组成了这所房子的结构。围绕下层的平板和墙是混凝土制成的,而最上层是带有木框架填充的复合钢和重型木结构。整体结构由杉木板制成的外墙覆盖,覆盖在传统的屋顶和墙体上。底部混凝土结构的热质量能够在冬天和夏天抑制房子里的温度浮动。在夏天,房子可以进行自然冷却,自然通风,新风从房子北面最低层进入,从中央裂缝上方排出。

Hadaway House

This snow country house is located on a northwest slope overlooking a panoramic view of Whistler Valley in southwest British Columbia. The site is a difficult wedged shape which offers just enough room for a garage and narrow entrance on the street side at the top of the slope.

The exterior form of the house is shaped by the intersection of two principal considerations: The first is the allowable building footprint and height. The second is the need to shed snow from the roof into appropriate storage areas within the site. The sculptural volume that emerges from this intersection of extruded building footprint and folding snow-shed roof is occupied in a manner that maintains the inherent plastic properties of the building form.

The main level is essentially one large space with living, dining and kitchen areas and an outdoor deck all of which open up to the valley view. A vertical crevice of space runs under the high-

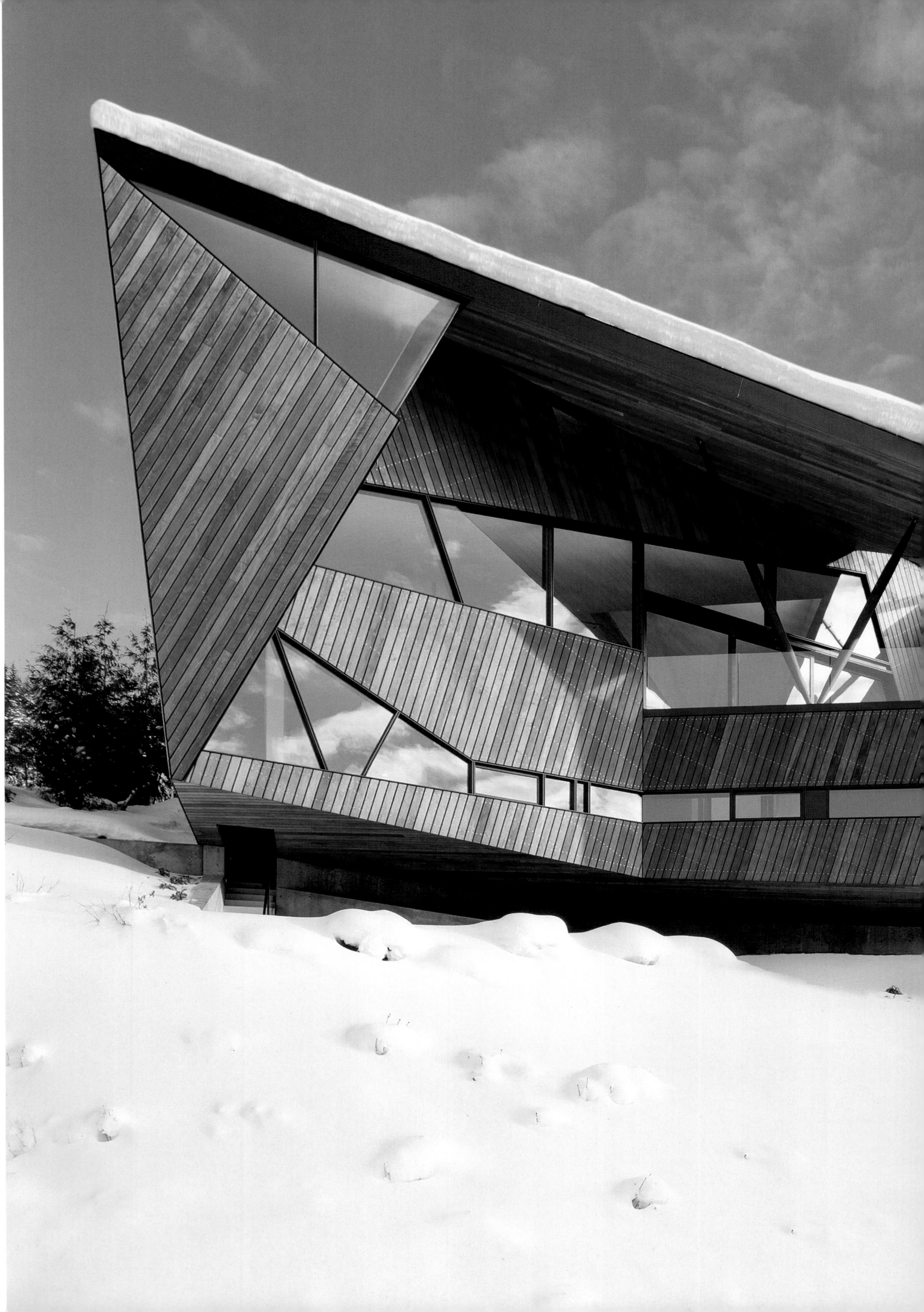

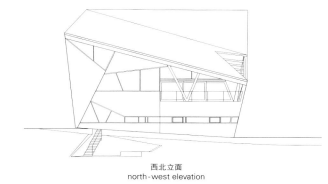

西北立面
north-west elevation

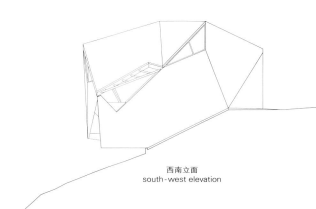

西南立面
south-west elevation

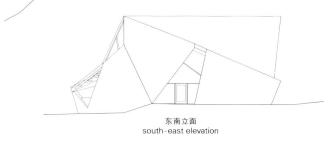

东南立面
south-east elevation

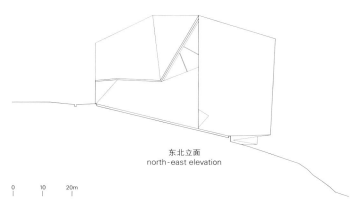

东北立面
north-east elevation

est roof ridge, bisecting the warped volume and bringing light to the deepest part of the section and plan. Stairs rise within this rift and a bridge crosses it at the upper level connecting master bedroom suite and study. Below, on the lowest level, are more intimate spaces housing guest bedrooms and a second living area, as well as a large service space. Accessible directly from the garage entrance to the house, this service space supports life in snow country – where wet clothes are hung to dry or thrown directly into the laundry, where skiers can store all the paraphernalia of their daily activities. Another stair connects this lower level to an outdoor patio below the house, the only ground level on the steep site other than that at the front entrance.

Construction is a hybrid of monolithic and framed systems. The slabs and walls which enclose the lower floor are concrete, while the uppermost levels are a composite steel and heavy timber structure with wood-frame infill. The entire structure is sheathed with a monolithic screen of open-spaced cedar boards over conventional roof and wall assemblies. The thermal mass of the lower concrete structure dampens temperature swings within the house in summer and winter. In summer the interior is naturally cooled and ventilated by drawing air from the lowest level on the north side of the house to vent at the top of the central rift.

项目名称：Hadaway House
地点：Whistler, British Columbia, Canada
建筑师：Patkau Architects
项目团队：John Patkau, Patricia Patkau with Lawrence Grigg, Stephanie Coleridge, Marc Holland, Peter Suter, Shane O'Neil, Mike Green
结构工程师：Equilibrium Consulting Inc. / 外围护结构：Spratt Emanuel Engineering Ltd.
地质技术：Horizon Engineering / 承包商：Alta Lake Lumber Co.
用地面积：1045m² / 建筑面积：299m² / 有效楼层面积：465m²
设计时间：2007 / 竣工时间：2012
摄影师：©James Dow (courtesy of the architect)

1 起居室
2 卧室
3 套间
4 大厅
5 滑雪设备存放空间
6 酒窖
7 浴室
8 储藏室
9 机械室

1. living room
2. bedroom
3. en-suite
4. hall
5. apres-ski
6. wine cellar
7. bathroom
8. storage
9. mechanical

一层 first floor

1 起居室
2 入口
3 门厅
4 餐厅
5 厨房
6 平台
7 车库
8 盥洗室

1. living room
2. entry
3. foyer
4. dining room
5. kitchen
6. deck
7. garage
8. powder room

二层 second floor

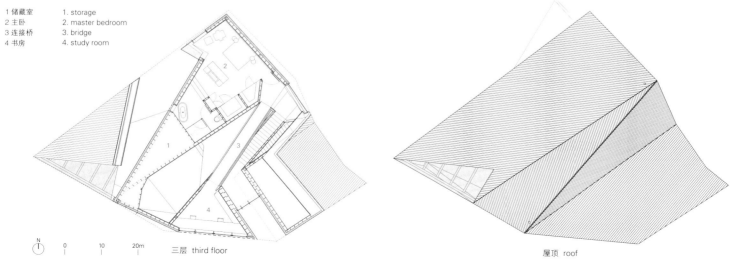

1 储藏室	1. storage
2 主卧	2. master bedroom
3 连接桥	3. bridge
4 书房	4. study room

三层 third floor

屋顶 roof

1. entry canopy assembly
- 1"x5.25" IPE wood cladding layer @6" O.C. fastened W/ 2" #10 ss truss head machine screws C/W washer and nut, through oversized pre-drilled holes to 3.5" 20 gauge galvanized Z-girts @24" O.C. (Z-girts C/W 18 gauge diagonal strap bracing members) fastened W/3/4" #14 butyl washered roofing screws to 1/2" wide folded top seam of 1.5" high 24 gauge standing seam prefinished metal roof
- ventilated underlayment membrane on roofing membrane
- 1/2" ply sheathing over 3/4" ply sheathing
- lvl (laminated veneer lumber) roof joists
- suspended wood framing
- 1/4" prefinished sheet aluminum soffit on 1/2" ply sheathing

2. typical roof assembly
- 1"x5.25" IPE wood cladding layer @6" O.C. fastened W/2" #10 ss truss head machine screws C/W washer and nut, through oversized pre-drilled holes to 3.5" 20 gauge galvanized Z-girts @24" O.C. (Z-girts C/W 18 gauge diagonal strap bracing members) fastened W/3/4" #14 butyl washered roofing screws to 1/2" wide folded top seam of 1.5" high 24 gauge standing seam prefinished metal roof
- ventilated underlayment membrane on roofing membrane

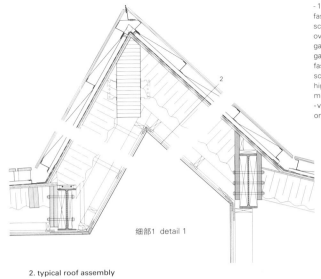

细部1 detail 1

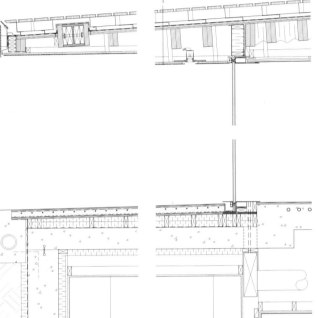

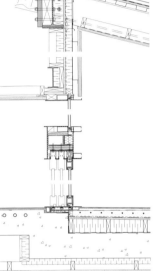

2. typical roof assembly
- 3/4" PLY sheathing
- 1.5" steel deck
- 10" structural steel framing
- suspended 2 x 4 framing
- continuous insect screen
- 1"x5.25" IPE wood soffit cladding layer

3. typical north canopy assembly
- 1"x5.25" IPE wood cladding layer @6" O.C. fastened W/2" #10 ss truss head machine screws C/W washer and nut, through oversized pre-drilled holes to 3.5" 20 gauge galvanized Z-girts @24" O.C. (Z-girts C/W 18 gauge diagonal strap bracing members) fastened W/3/4" #14 butyl washered roofing screws to 1/2" wide folded top seam of 1.5" high 24 gauge standing seam prefinished metal roof
- ventilated underlayment membrane on 2 PLY SBS roofing membrane
- 3/4" PLY sheathing
- 1.5" steel deck
- 10" structural steel framing
- suspended 2x4 framing
- continuous insect screen
- 1"x5.25" IPE wood soffit cladding layer

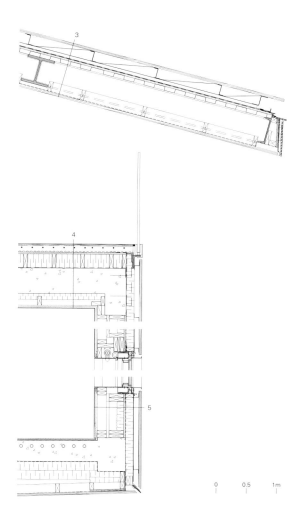

4. typical deck assembly
- 3/8" tile
- 3/8" sanded grout
- latex thin-set mortar additive
- crack suppression membrane
- 1.5" thick bed mortar mix W/electric snowmelt system
- drainage mat
- SBS waterproofing membrane
- 3/4" PLY
- strapping W/1:50 fall to drain C/W R20 rigid insulation
- self adhesive air/vapor barrier membrane
- concrete slab

5. typical wall assembly
- 1" IPE wood cladding layer on
- 2" horizontal slotted galvanized Z-girts
- 26 gauge vertically shingled PRE. FIN. galvanized metal liner
- 2.5" R12/5 rigid insulation C/W vertical galvanized Z-girts
- self adhesive air vapor barrier
- 3/4" plywood sheathing
- 2x6" framing C/W 2.75" R8 spray foam insulation
- 2x4 strapping
- 2 layers 1/2" GWB

1 连接桥 2 储藏室 3 入口 4 门厅 5 起居室 6 大厅 7 浴室 8 套间
1. bridge 2. storage 3. entry 4. foyer 5. living room 6. hall 7. bathroom 8. en-suite
A-A' 剖面图 section A-A'

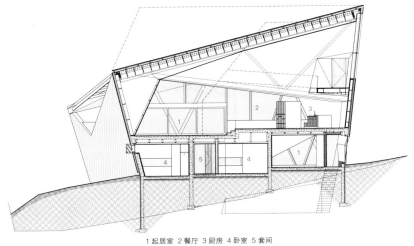

1 起居室 2 餐厅 3 厨房 4 卧室 5 套间
1. living room 2. dining room 3. kitchen 4. bedroom 5. en-suite
B-B' 剖面图 section B-B'

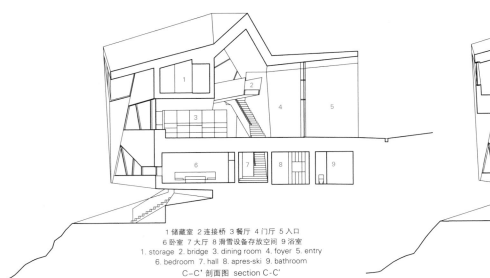

1 储藏室 2 连接桥 3 餐厅 4 门厅 5 入口
6 卧室 7 大厅 8 滑雪设备存放空间 9 浴室
1. storage 2. bridge 3. dining room 4. foyer 5. entry
6. bedroom 7. hall 8. apres-ski 9. bathroom
C-C' 剖面图 section C-C'

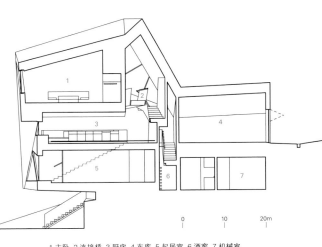

1 主卧 2 连接桥 3 厨房 4 车库 5 起居室 6 酒窖 7 机械室
1. master bedroom 2. bridge 3. kitchen 4. garage
5. living room 6. wine cellar 7. mechanical
D-D' 剖面图 section D-D'

四季住宅
Casos de Casas

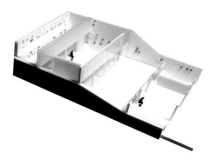

设计微气候以提高家庭生活舒适度

本案住宅结合了一种特殊的城市规划形式,这种形式在西班牙比纳罗斯以及卡斯特利翁海岸地区某些地方常见。比纳罗斯开发出了一种紧凑的、适当的规划形式,以适应当地一年中主要的两个季节。一个地方两种生活方式:一种是正式的围绕着城市的生活方式,而另一种是非正式的享乐主义生活方式。

在城市中心,住宅塔楼越来越多,水平的城市消除了海边多样的土地划分,创造了一个没有街道和公共空间的城市,那些被密集使用的老道路现在都铺设了铺地材料,供车辆通行。在这个背景下,此住宅成为比纳罗斯市民在炎热季节的避暑之处,或者国内游客的第二居所,展现了具有两种不同居住设施的一座城市。

提升家庭生活舒适度

客户的日常生活被航班、城市、工作、行程这类事项排满了,无法享受日常的居家生活。而该住宅,作为一个交叉点,构建了浓厚的家庭生活氛围。这是因为四季住宅没有固定的人居住。

该项目在一个大斜坡上建有一个大型运动场,从这里可以观赏到街道上看不到的房屋全貌。

这个运动场与周围景观融为一体。活动没有特定位置,根据场所的特点、视角、气候及私密性而设定,这使该项目成为应对各种家庭生活愿望和需求的最优化的设计方案。在运动场内,桌子和灯具还被设计成大型设施,供运动场内的人使用。

高度不同的两个顶层遮挡了运动场,将其与最私密的活动区连接起来,使其成为一个连续、无限制的环境。用餐在楼上,就寝在沙发上。可以在后院读书,还可以在楼前洗澡。

一座地中海式的全球性建筑

房屋如果没有波浪起伏的瓷砖及塑料材料,又怎么能算作是"全球性"的呢?上千人曾在这个功能性大伞下面居住过。这个系统可以是非正式的,甚至更加复杂高效。四季住宅还是一座没有固定场所的住宅,这里没有地中海风格的装置。白色的外观能有效应对太阳辐射与气候,这也使该建筑成为当地住宅中的一个异类。它设计有标准化及可调节的结构体系,满足了房主个性化的需求。建筑由内而外建造,为适应不同的气候类型,不同侧面的双层墙体分别采用了多孔聚碳酸酯、波状聚碳酸酯简易滤光片、锁扣金属瓷砖以及不同的玻璃。农工织物能够消除或积聚热能,这是一处既开放又封闭的住宅。其透明部分看似封闭,而其封闭空间却又看似开放。房屋有一个"等待结构",能够满足客户对于更高水平节能的构想。

Seasonless House

Design your Climate to Increase Your Domesticity

The house combines a particular form of urban planning that is taking place in Vinaroz, and in some places on the coast of Castellón. The city has developed a compact and suitable form of urban planning, based simultaneously on the temporality of the two main local seasons of the year. There are two ways of life in one place. One organized around the urban and the formal, and the other organized in a more informal and hedonist environment. In the center, housing towers accumulate and the horizontal city eliminates the diversity of allotments in front of the sea, creating an urban plan without streets and public spaces derived from the intensive use of old roads, now paved for vehicles. In this context the hot season housing of the citizens of Vinaroz or the second homes of visitors from the interior, present a city with two different habitable infrastructures.

Increased Domesticity

The de-locating of the client's domesticity, which is spread across his practices: flights, cities, work, itinerary intimacies, doesn't per-

东立面 east elevation

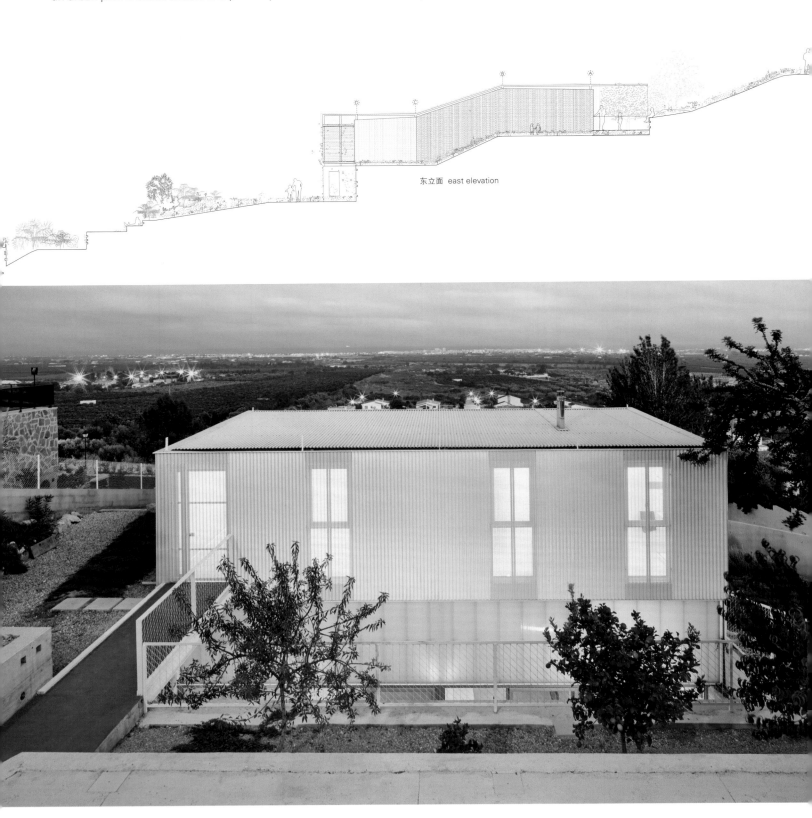

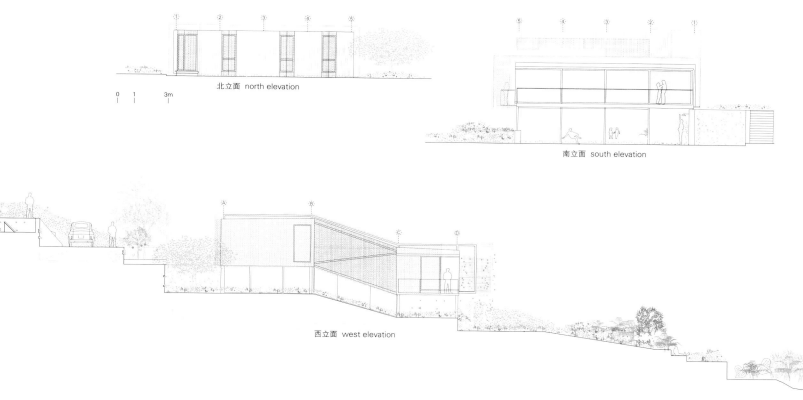

北立面 north elevation

南立面 south elevation

西立面 west elevation

mit the construction of a close and accessible day to day domesticity. The house, as a crossing point, builds an augmented domesticity. It is because of this, that the Seasonless House is not inhabited. In this system, the house builds a large playground from the strong slope of the land, a place from which to observe and include the house landscape, without being clearly observed from the street. It is in this playground where a more indeterminate and polyfunctional way to live with the landscape is provided. Activities do not have a specific location in which to develop but rather, according to the local aspects, views, climate and privacy, they enable the project to develop as a landscape of variable domestic events that respond to desire, need or optimization. It is in this place where the "far away table" or the "squid searching lamps" are designed as large-scale devices for interactions with the playground.

The top floor on two different heights covers the playground connecting it with the most private activities. Reality surpasses both in one continuous and limitless environment. Eating happens upstairs and sleeping on the sofa. Reading takes place in the backyard and one can shower at the front of the building.

A Mediterranean Global Construction

What is "globality" if not a house of undulating tiles and plastic? Thousands of people live under this fundamental umbrella. But this system can be very informal or more sophisticated and efficient. The Seasonless House is also a house without a place, a Mediterranean styled device. White, efficient for climate and solar radiation reasons is also efficient as a singular element of the roaming nature of its inhabitant. Its execution: a standardized and modulated construction system, personalized for specific needs in this house. Built from the inside out, with double walls made from cellular polycarbonate, simple filters of undulated polycarbonate, interlocked metallic tiles, different glasses according to the contiguity of the climate on either of its sides. The agro-textiles dissipate or accumulate the radiation. An open and closed house at the same time. That transpires where it appears closed and is hermetic where it appears open. The house has a "waiting structure" in preparation for when the client attains a greater level of energy efficiency.

二层 second floor

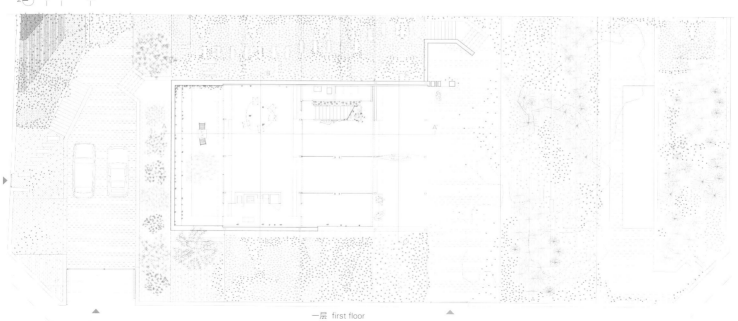

一层 first floor

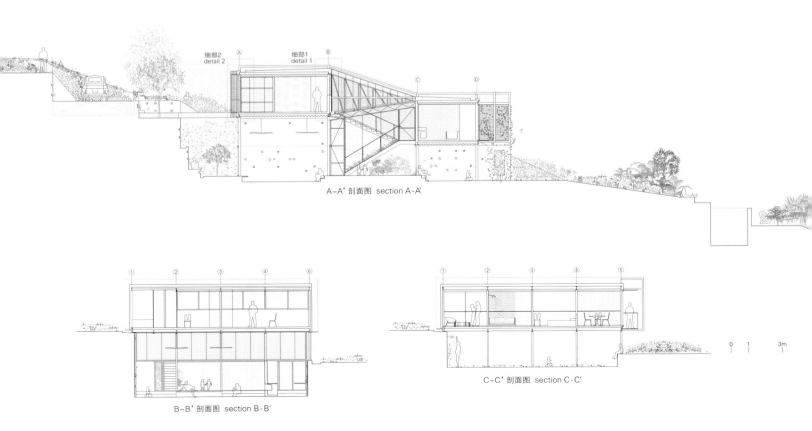

A-A' 剖面图 section A-A'

B-B' 剖面图 section B-B'

C-C' 剖面图 section C-C'

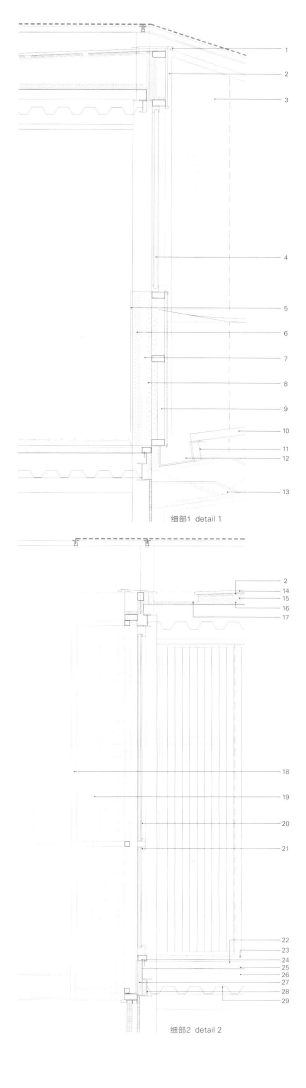

细部1 detail 1

细部2 detail 2

项目名称：Seasonless House
地点：Vinaroz, Castellón, Spain
建筑师：Irene Castrillo Carreira, Mauro Gil-Fournier Esquerra
合作者：Mª Eugenia castrillo
技术建筑师：project_José María Herás/
work_Cesar Villalonga. Cota Zero
结构工程师：Francisco Fiol. Fitconsult, sl.
占地面积：36129m²
设计时间：2011 / 竣工时间：2013
摄影师：©José Hévia(courtesy of the architect)

1. stainless steel sheet lac 9003 e: 1m
2. undulated steel sheet lac 9003
3. white cellular polycarbonate modulit 500/40mm
4. sun energy multilayer glass control solar 4+4/8/4
5. interior pladur e: 15mm
6. rockwood panel e: 60mm
7. pladur double layer hidrofugal e: 15+15mm
8. insulation polycarbonate vapor layer e: 80mm
9. ventilated air camera e: 60mm
10. transparent undulated polycarbonate SINUS 177/51
11. OZ stainless steel frame 15-5cm
12. steel sheet frame e: 1mm
13. tubular steel frame
14. stainless steel substructure 60×40
15. inclined mortar+geotextile
16. rigid insulation 70mm
17. stainless steel sheet e: 1.5mm
18. window steel frame 40×40
19. micro perforated undulated steel sheet lac 9003
20. multilayer glass stadip 4+4/8/4+4
21. steel window frame
22. cement nivelation from wood floor
23. wood floor
24. steel tube 80×40
25. rigid insulation radiant floor e: 60mm
26. steel structural steel compression concrete layer 6+14
27. flexible insulation e: 60mm
28. steel frame IPE 140
29. stainless steel sheet HLM 60/220 e: 1m

自然与人工——一分为二还是合而为一?

Natural and Artificial: Dichotomy or Duality?

如果，我们首先思考印度哲学，就能够接受这种可能性：自然与人工完全是一回事——因为人类是大自然的一部分，并且人类所有的行为都不能超越自然法则。

将自然与人工这两个条件对立是另一种思考这个课题的方式。根据这一思路，我们可以将建筑学定义为一种人工活动——因为它是人类智慧的结晶和改造自然的行为，有一种重新调整结构逻辑的能力。

但是我们也能根据其设计方法来分析建筑产品，将它定义为自然的或人工的事物——以一种具体的方式来考虑一定的规则、承袭文化遗产、与环境和谐相处、选择一种不同的过程、在施工中使用适当的技术，以及明显使用自然元素或人工制品等。

在接下来的项目中，我们将尝试去了解建筑师怎样通过处理自然与人工这两个条件来实现一个完整的、和谐的结果，还将了解他们怎么创造一个功能型的建筑物，在自然和人工之间维持好平衡。

If, as a depart, we think about Indian philosophy, we can accept the possibility that Natural and Artificial are one and same – as Humans are part of Nature and all human action can not go beyond the very laws of Nature.

Another way of considering the subject is by opposing the two conditions: Natural vs. Artificial. Following this line thought, we can define Architecture as an artificial activity – as it is a result of human's intelligence and transformative action over Nature, being a capacity of reorganizing its structure onto a new logic.

But we can also analyze the architectonic production according to its approach, defining it as a Natural or an Artificial one – in its specific way of respecting certain rules; of following cultural heritages; of harmonizing with the surroundings; of opting on different processes; of using suitable technologies in the construction; of making an evident use of natural elements or artifices; etc.

In the following projects, we try to understand how architects manage the two conditions, Natural and Artificial, to achieve a whole harmonic result and how they create a functional architecture and construction, balancing between Nature and artifice.

Pacific Dazzle美发沙龙_Pacific Dazzle Baton/Atelier KUU
欢乐餐厅_Cheering Restaurant/H&P Architects
紫禁城红墙茶室_The Forbidden City Red-wall Teahouse/CutscapeArchitecture
Steirereck餐厅_Restaurant Steirereck/PPAG Architects
Angolino餐厅_Angolino Restaurant/Geneto
自然与人工——一分为二还是合而为一？_Natural and Artificial - Dichotomy or Duality?/Paula Melâneo

随着文明的发展，人类离自然越来越远。城市扩张导致形成庞大的城市环境，必然使人类与自然远离。在这个框架下，我们到达了一个不再讨论自然风景的地步，而是谈论领土和人们管理占地的方式。

在过去的几十年里，我们注意到对这一情形的担忧正在增加。很多人开始寻求与自然和谐相处的新方法。一些人彻底搬到市区以外的地区，而其他人依然在人工的环境下尝试使用人工方法发展与大自然相结合的新体制，或是从生态学的视角通过探索新科技来引入自然元素。

实际上在21世纪，人类的科技发展取得了巨大的进步，人们在日常生活中频繁地使用技术，技术被认为是一种优秀的人造工具。在这种视角下，正如法国社会学家和人类学家布鲁诺·拉图尔所说的："为了与自然相处，我们不得不接受人造事物，这有些自相矛盾。"实际上，技术是恢复和平衡人与自然的关系的有效（人工）手段，无论是从怀旧还是改革的角度来看都是如此——通过发展科技，人类有更多的可能去发现自然现象。

现在，在一个经济不对称和分裂的信息时代，人类依然不断地在发明新的工具和机械装备来支配自然。最后，由于人类始终在尝试人为地构建自然，导致了人工产品被附加于原本是自然的事物之上的情况，比如景观设计、生态学、生活、生态系统、智能产品等。

自然与人工的主题很大，而且是主观的。不过我们仍然可以指出，普遍的偏见包含这两种情况：它们被一分为二，自然意味着积极的观点，而人工与消极的事物联系在一起。然而，自然与人工之间的界限却越发模糊，尤其是在建筑学，促成了那个死板的二分法。

物质性的新概念正在出现，正远离建筑学。技术是改革的决定性因素。计算机辅助设计（CAD）和参数生成的建筑设计正在改变自然结构的范例和人们对它的理解——甚至是复制自然结构。材料研究是一个不断变化的领域，人工产品在这里通常是对使用或滥用自然材料的肯定的回答，而从生态学的角度来看，资源可能是一种危险的回应。

在这里报道的项目基本都是拥有普通功能的建筑物，可以容纳零售活动（一家茶馆、三家餐厅和一家美发沙龙）。其设计使用了适宜的人工和自然构件尺寸，从而有效地安置了这些空间。

As civilization developed, humanity got more and more distant to nature. Increasing urban growth led to large city contexts that necessarily detached man from nature. Within this framework, we arrived at a state where we no longer talk about natural landscape, but about territories and the way man manages the land occupation.

In the last few decades we observed a rising concern about this situation. A new path of reconciliation with nature started to be a desire of many. Some radically moved to non-urban areas, others are still trying to develop new systems of integrating nature in manmade environments using artificial means or, with an ecological perspective, introducing natural elements by exploring new technologies.

Actually man reached great technological advances in the 21st century, intensely using technology in daily life, which can be considered an artificial tool by excellence. In this perspective, as French sociologist and anthropologist Bruno Latour states, "Paradoxically, to cope with the natural we have to embrace the artificial." Technology is in fact a valid (artificial) device to retrieve and balance the relationship between man and nature, in a nostalgic or in an evolutionary sense – by developing technology, humanity opens the possibilities of discovering (or understanding) new natural phenomena.

Nowadays, in an informational era of economic asymmetries and disruptions, humanity is continuously inventing new tools and mechanisms to dominate nature. Ultimately, humanity is always trying to artificially construct nature, resulting in the condition artificial being used as a prefix for things that are natural at a very beginning, such as Landscapes, Ecologies, Life, Ecosystems, Intelligence, etc.

The theme of Natural and Artificial is vast and subjective. Still, we can denote that a general prejudice encompasses these two conditions: they are seen as a dichotomy in which Natural signifies a positive perspective and Artificial is associated with a negative point of view. However, the boundaries between Natural and Artificial are increasingly blurring, particularly in architecture, contributing to review on that rigid dichotomous idea.

New concepts of materiality are emerging, moving away from the tectonic. Technology is a definitive factor for transformation; computer aided design (CAD) and parametric generated architecture are changing paradigms and the understanding – and even reproduction – of natural structures. Materials' research is an ever-changing field, where "artificial" is frequently a positive answer as the (ab)use of natural components and resources can be a dangerous response, from the ecological point of view.

The featured projects are based on mundane programs, which

紫禁城红墙茶室，中国北京
The Forbidden City Red-wall Teahouse in Beijing, China

Pacific Dazzle美发沙龙位于神户一条繁华的街道上，日本建筑设计公司Atelier KUU想创造亚洲花园和山峰般的平静和安宁。在这里，舒缓的气氛是设计的基本理念，为此我们可以说他们的设计方法是贴近自然的。所以许多自然元素都被加了进来，比如植物、小树、绿茵，甚至是小喷泉里流淌的水。但是，由于使用了许多人工产品，所以景色也是人工的——主入口的立面是一面墙，墙上切了不同高度的盒子结构，做成花盆，里面栽着植物，重新诠释了多岩石的表面；在入口处，小喷泉添加了潺潺的流水声；隔离屏风带有植物图案；用一组细杆作分隔带，让人想起一片竹子；苔藓覆盖了一部分室内的几何图案表面；起居室里的感觉就像在树荫下；工作环境中的光影给人感觉就像身处乡村的艳阳天里——这些都是为了获得孤独与安宁的感觉，就像我们能在大自然中找到的一样。

该项目力求在人工与自然之间寻找一种平衡，并以某种方式联系日本文化，其中小规模重建这一自然的想法是日常生活中的一部分。

紫禁城红墙茶室由中国的建筑设计工作室研造社设计而成，将两个仓库改造成了新的茶室。这些老建筑位于紫禁城外墙附近的一条胡同（传统的中国/北京街道）里，紫禁城的外墙刚好在这里局部遭到毁坏，所以直接连着皇家园林。建筑师没有选用传统的中国茶室的设计，而是采用了一种现代方法，同时还考虑了两种不同的体验自然的方法：胡同居民每天的日常生活，以及该地的文化遗产——关于这个地方的回忆，这里以前是皇室纪念先祖的庙宇，有皇宫和皇家花园等历史遗产。这个过程可以说是很自然的，考虑到了对从前的记忆和遗迹的保存。

两个仓库中的一个连接着皇宫外墙，就保留了外层砖墙的原样，其中容纳了一个新的钢结构框架，结构框架中有几个独立的体量，用于几个隔离开的私人茶室。在这些新体量和以前就有的墙之间，小院子放大了外部空间，并且平衡了新旧的关系。

另一个仓库保留了主要结构，有三个茶室以及一个更具功能性的空间。所有的茶室都有与众不同的特性，鲜明地表现在了各种不同覆层材料的使用上。在这里使用了自然材料和人造材料（木头、石头、竹子、钢、玻璃、铜、砖块），在每一个空间都能产生不同的体验。

allow to house retail activities (a teahouse, three restaurants and a hair salon). It is design and architecture that gives them the successful direction, by using an appropriate measure of Artificial and Natural components in their approach.

In the case of the Pacific Dazzle Baton, a hair salon placed in a Kobe city's busy street, the Japanese practice Atelier KUU thought about recreating the quietness and calmness of an Asian garden or mountain. Here, the soothing ambience was the basic idea for what we can call a natural approach on their design. For that, natural elements were included, such as plants, small trees and greenery and even water in a small fountain. But the scenario is artificial as using several artifices – the main entrance facade is a wall with carved boxes where the plants are inserted at different levels, which can be a reinterpretation of rocky surface; the small fountain, at the entrance, adding the sound of trickling water; the separating screens with floral motifs; partitions with a set of sticks reminiscent of a group of bamboos; mosses cover some interior geometric surfaces; the living room giving the sensation of being under the shade of a tree; working light and shadows as they can exist in a sunny day in the countryside – to achieve that sense of solitude and tranquility, as we can find in Nature.

This project is seeking for a balance between the two conditions, artificial and natural, somehow related with the Japanese culture, where miniature reconstructions of an idea of nature are part of the common daily life.

The Forbidden City Red-wall Teahouse, designed by the Chinese office CutscapeArchitecture, consists of the renovation of two warehouses into a new teahouse. Those old buildings were situated inside a Hutong (typical Chinese/Beijing neighborhood) just beside the exterior wall that surrounded the Forbidden City, in the exact place it was partially demolished and, for that, connecting directly with the royal garden. Keeping away from the traditional design of the Chinese teahouse, the architects adopted a contemporary approach while respecting the two different life experience natures of the place: the everyday life of the Hutong's inhabitants and the cultural legacy of the place – concerning the memory of the site that served previously as a royal memorial temple for ancestors, the historical heritage of the imperial palace and the royal garden. This procedure can be considered as natural, taking into account the preservation of the previous times' memory and remnants.

One of the warehouses, attached to the palace wall, just keeps the exterior brick walls and houses a new steel-frame structure of individual volumes for several differentiated and private tearooms. In between the new volumes and the pre-existing walls, small courtyards amplify the interior space and balance the rela-

欢乐餐厅，越南河内
Cheering Restaurant in Hanoi, Vietnam

这个项目的所有组成元素都对景观格局特别关照，因此我们无法清晰地定义花园的尽头在哪里，也不知道茶室从哪里开始，为这个空间营造了一种崭新的自然环境。

越南的H&P建筑师事务所在河内市中心设计了欢乐餐厅。繁忙的人行道是该地区的特色，行人经常在古老的树下吃饭，这些古树点缀于城市肌理中。建筑师人工重建了河内城市的"自然环境"及其餐厅中的文化——新型木结构占据了已有的玻璃和钢材外层。这种新型木质模型介于传统亚洲木质建筑和具体化的小树林之间。该结构大部分为人造的，好似由一种非结构木梁搭成的积木，包括主要的基础设施，扮演了断层块木梁的角色。它们彼此垂直，并在空间中垂直分布，在透明的聚碳酸酯屋顶下，创造了一处活力空间，产生了不同的光影效果——比如从真正的树枝中穿透过来的光线。一棵古树——在原先基地上剩下的自然元素——在项目设计中经过考虑被保留下来，占据了内部空间，并用木质模型在整个休息厅内到处复制古树的外形。雨水系统带来凉爽之感，并净化了空气，它是一种控制高温的自然方式。

在某种程度上，该设计可以被看成一种自然的方法，因为这家餐厅努力再现了传统城市生活和自然的精神，营造了一个流动的空间，并将之转变为街道的延续。

另一家餐厅的设计则使用了完全不同的解决办法。在意大利餐厅Angolino中，日本设计工作室Geneto明显采用了一种人工的设计，来回应甲方和当地环境的需求：在附近地区全新的建筑外形，一个紧挨着户外风景的自建体量。甲方希望这家餐厅能带给馆林市一种新的身份，像标志性建筑物那样产生重要的影响。这件人工作品将在建筑物和当地喜欢来这个愉快的地方聚会的人们之间创造一种关系——这是一种应当在施工中就开始的关系。作为一座自建的建筑，在工程期间它激发了与当地行人的交流，让他们与餐厅的业主对话，就这里将要建造什么样的建筑交换意见。同时，它也应当成为保证餐厅顾客隐私的封闭空间。因此，设计出了一个有着许多棱角、尖屋顶——模仿了背景中关东地区山脉的轮廓——以及小洞口的硬壳式结构。它让人想起了舒适的住所，因为能够创造情感上的亲密关系，所以可以将它看成是一种自然的设计方

tion between new and old.

The other warehouse keeps the main structure and contains three of the tearooms and a more functional part of the program. All the tearooms have distinctive natures, strongly marked by the use of different cladding materials. Natural and synthetic materials are applied (wood, stone, bamboo, steel, glass, copper, brick) and allow experimenting different sensations in each space.

The whole assembly of this project has a special care on the landscape arrangements, so we cannot clearly define where the garden ends and where the construction of the teahouse begins, creating a new nature for this space.

The Vietnamese practice H&P Architects designed the Cheering Restaurant in Hanoi's center. Lively sidewalks characterize this city's area, where people often take their meals under the ancient trees, which punctuate the urban fabric. The architects artificially recreated Hanoi's city urban "natural environment" and culture in their restaurant – a new wooden structure that occupies an existing outer shell of glass and steel. This new wooden matrix lies somewhere between the traditional Asian wooden architecture and the materialization of a small forest. Assumedly artificial, a game of wood-covered non-structural beams, containing the main infrastructures, plays the role of timber massif beams. Piled perpendicularly to each other and vertically distributed in the space, under the transparency of the polycarbonate roof, they create a dynamics where different shadows occur – such as the light passing through the branches of real trees. An old tree – a natural element as a remnant of the original site – is taken into consideration in the project and occupies the inner space, replicated throughout the lounge by the wooden matrix. A rainwater system is previewed for cooling and cleaning the air, as a natural mean to control the heat.

In a way this can be considered a natural approach as this restaurant tries to recreate the spirit of the traditional city life and nature, creating a fluid space that is transformed into the continuity of the street.

Another typical program for a restaurant is solved in a completely different way. For the Italian restaurant Angolino, the Japanese studio Geneto designed clearly an artificial response to the particular needs of their client and the local context: a completely new shape of building in the neighborhood, a closed-to-the-exterior-views volume that would be self-constructed. The client wanted this restaurant to give a new identity to the Tatebayashi city, making a difference as a landmark. This artifice would create a relationship between the building and the locals who would see

Steirereck餐厅，奥地利维也纳
Restaurant Steirereck in Vienna, Austria

法。胶合板被选为主要的材料，因为它容易获得，也容易在自建的门式框架结构系统及其覆层中使用。在内部它是可见的，强调了自然之感和热情好客的态度。安装了三角形小木窗的墙体厚度不大，促进了室内外的空气流通，也没有影响内部舒适安逸的气氛。以上所有方面都将被自然而然地结合到这座位于社区内的建筑中，成为附近居民日常生活的一部分。

Steirereck餐厅位于维也纳的城市公园中。在其扩建设计中，PPAG的建筑师想要在这个绿意环抱的公园的特殊环境中创造一个新式的就餐环境。由此诞生了一个片段式的反射结构，建筑师打算在其中结合自然材料和高科技材料。几个体量被设计得好像是几个小亭子，它们彼此相连，为就餐提供了独立而舒适的空间。在这里，连接室内外的地方无论在视觉上还是实体上都显得十分特别，这都是因为那些巨大的电动玻璃窗。由玻璃和金属包裹的立面反射了公园里的绿色环境，产生了艺术品的效果，让新建的结构几乎消失在环境中，强调了自然的成分。在外部，小院子和露台与内部空间融合在一起，仿佛延续了内部空间。这个新建筑上方为绿色屋顶，为烹饪提供了药草和食材原料，也将建筑隐藏起来，让自然赋予与人工创作和谐相处。对于原有的结构，在就餐休息室中，建筑师在内部采用了与新的反射立面相同的金属材料，创造了一种与外部空间的对话，就像他们在内部创造了一种自然气息。天花板被转变成了一种有机形状，让人联想起了地形分析的等高线。与此相反，餐厅的中间部分以瓷砖覆盖表面，作为对天花板花纹的计算机算法的回应，参照了厨房的设计，算是一种人工装置。

the restaurant as a pleasant space for gathering – a relationship that should be initiated during its construction. Being a self-built structure, it activates the communication with the locals passing by during the works, allowing them to dialogue with the owner, changing opinions about what would arise there. At the same time it should be a closed space to guarantee the privacy of the restaurant's clients. For that, a monocoque volume with strong angles, pointed roof – as mimicking the skyline of the Kanto Mountains in the background – and small openings was designed. It suggests the feeling of a cozy shelter, which can be considered as a natural approach, as being able of creating emotional affinities. The plywood was chosen as the main material, because it is easy to get and to use in this self-built portal frame structure system and in its covering. It is visible in the interior, emphasizing that natural and welcoming feeling. The low thickness of the walls with the small triangular windows allows an exchange of atmospheres between interior and exterior, without compromising an interior cozy ambience. All these aspects would naturally integrate the building in the community as part of their daily life.

The restaurant Steirereck is located at the Stadtpark (City Park) in Vienna. For its extension, PPAG architects had in mind to create an innovative typology for a dining space, within the exceptional context of the green surroundings of the park. The result was a fragmented and reflective structure, where they intended to combine natural and high-tech materials. Several volumes were designed – as if they were small pavilions – and interconnected, providing independent cozy spaces for dining. Here, the link interior/exterior is privileged, both visually and physically, by the large electric sash windows. The glass and the metal-covered facades generate an artifice for the reflection of the green atmosphere of the park, dematerializing the new intervention and emphasizing the natural component. In the exterior, small courtyards and terraces merge with the interior in a spatial continuum. This new construction is covered with a green roof, which supplies herbs and ingredients for the cuisine and work as a camouflage for the building, harmonizing the whole – natural and artificial. For the existing structure, in the dining lounge, the architects brought the same metallic material of the new reflective facades to the interior, creating a dialogue with the exterior space, like they were bringing nature inside. The ceiling was transformed in an organic shape, reminding the contour lines of terrain morphology. In contrast, in the middle section of the restaurant, the surfaces are covered with tiles responding to an algorithmic pattern, a reference to a kitchen, as an artificial device. Paula Melâneo

Pacific Dazzle美发沙龙

Atelier KUU

建筑师在对一间出租房进行翻新和改造时,首先在更深层次上了解了目前的空间特色和位置。

建筑师对一些还可以利用的元素进行了重构。这个出租房是一处80m²的开放空间,天花板高6m。建筑师运用多种可能性和理念将这处活力四射的开放区域的面积最大化。

建筑师设计的这座Pacific Dazzle美发沙龙是神户的第三个分店,Pacific Dazzle美发沙龙在大阪开了三家店,在神户开了两家店。它们旨在成为连不喜欢去美发沙龙的人都愿意去休闲和放松的地方。建筑师认为植物和水是必要的元素。

设计概念是"Shityu-Sankyo"。它是最有影响力的茶道学校"Urasenke"的语言。

它表明了人们在日常生活中也可以领略和享受这处非同寻常的空

间，且在城镇中也能欣赏到自然风光。

建筑师认为这个概念是联系这处空间和"Pacific Dazzle"美发沙龙的基础概念。

建筑布局简单明了，分布着多个盒子结构，每个结构设有一种功能。

建筑师努力使绿色不成为一种表达"Shityu-Sankyo"风格的装饰品。

种植在盒子结构上的树木围绕着整处空间，即使在室内人们也可以感受到树下的荫凉。

从墙体上放置的保鲜花开始，人们便能看出自等候大厅开始的流线形设计，不管在哪个地方，都能感受到绿色的气息。

建筑师认为把美发沙龙设计成令人舒适且不同寻常的空间无论在"Urasenke"学校期间还是现在都是一样受人欢迎的。

Pacific Dazzle Baton

When we deal with the refurbishment and modification of a rental, we understand the present individuality and situation of space on a deeper level.

It is for reconstructing the element which can be harnessed to the thing more than now.

This rental is the open space of the floor area of 80 m^2, and ceiling height of 6 m. We felt much possibility and thought to maximize this dynamic open space.

We designed the third store in Kobe of the beauty salon "Pacific Dazzle" which develops three stores in Osaka and two stores in

Kobe. They aim at making the space in which a beauty salon can be enjoyed and relaxed for everyone who doesn't like to go to a salon. We thought the green and water is the one of conditions.

The concept is "Shityu-Sankyo". It is the language of the most influential tea ceremony school, "Urasenke".

It is the implication of taking in and enjoying unusual space in the usual life and denotes that a natural sight can be felt in a town.

We believed this concept became a fundamental concept which connected with this space Pacific Dazzle.

The layout is a simple composition which sprinkled each function as a box.

We tried to design the green not to be an ornament for expressing "Shityu-Sankyo" style.

The trees on a box wrap the whole space and people can feel the shade of a tree even indoor.

Setting out from the preserved flower on the wall which can be conscious of flow lines from the waiting lobby people can feel green regardless of location.

We felt that the designing of a comfortable and unusual space for a beauty salon is all the same welcomed in the time of Urasenke and now. Atelier KUU

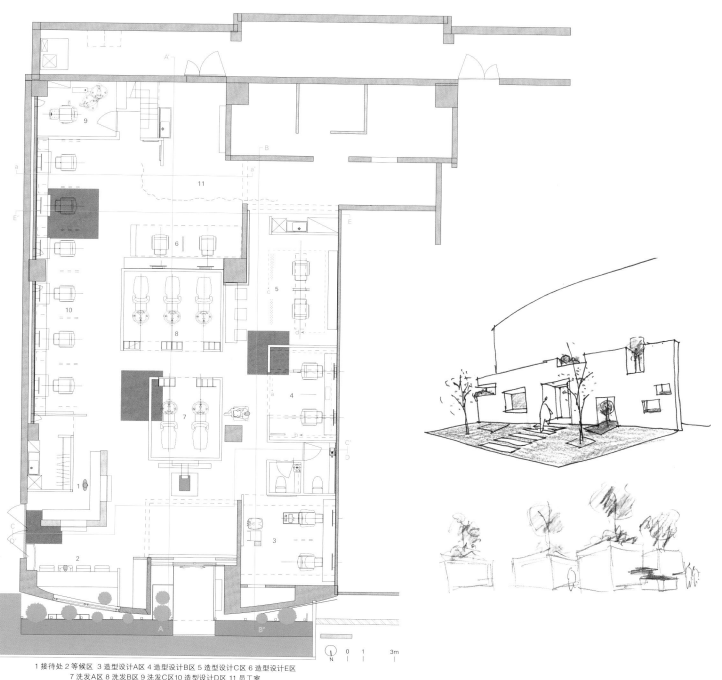

1 接待处 2 等候区 3 造型设计A区 4 造型设计B区 5 造型设计C区 6 造型设计E区
7 洗发A区 8 洗发B区 9 洗发C区 10 造型设计D区 11 员工室
1. reception 2. waiting space 3. styling A 4. styling B 5. styling C 6. styling E
7. shampoo A 8. shampoo B 9. shampoo C 10. styling D 11. staff room
一层 first floor

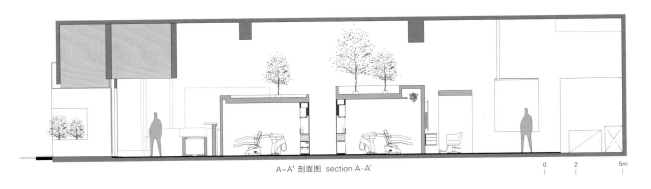

A-A' 剖面图 section A-A'

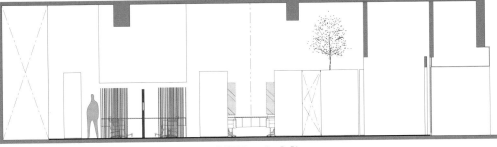

B-B' 剖面图 section B-B'

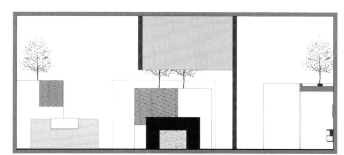

C-C' 剖面图 section C-C'

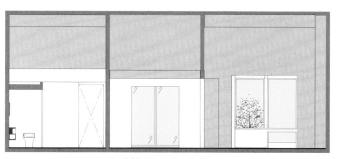

D-D' 剖面图 section D-D'

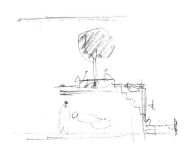

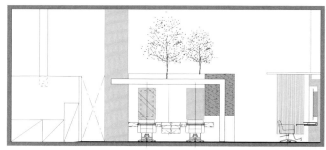

E-E' 剖面图 section E-E'

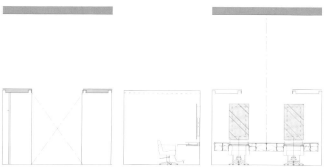

展开立面——造型设计B区（从a到b）
development elevation_styling B (from a to b)

展开立面——造型设计C区（从c到d）
development elevation_styling C (from c to d)

0 2 5m

项目名称：Pacific Dazzle Baton
地点：1400-51, Myodanicho, Tarumi-ku, Kobe-shi, Hyogo, Japan
设计师：Nobuo Kumazawa, Yuuki Kunimatsu, Harumi Oura
设计团队：Nobuo Kumazawa, Yuuki Kunimatsu_atelier KUU co.,ltd / Harumi Oura_HARE
甲方：Soichi Hanaki
建筑面积：253.2m²
材料：Floor_Metal trowel finished mortar / Wall_AEP coating, preserved moss / Ceiling_skeleton, AEP coating
竣工时间：2013.2
摄影师：©Seiryo Yamada (courtesy of the architect)

a-a'剖面图——造型设计D区
section a-a'_styling D

欢乐餐厅

H&P Architects

欢乐餐厅位于越南河内市中心，由一座长期废弃的项目翻新而来，它保留了钢框架结构和可重复利用的覆盖物，如玻璃、钢材、条钢、金属板屋顶。

河内居民的许多日常活动都是在街边人行道上进行的，尤其是一日三餐的烹饪。设计者从中汲取灵感，创造了一处以古树为主题的空间——也是这座千年古城街道的一幅熟悉景象。

该方案将整个原有钢材焊接成一个新的结构体系，内部敷设技术设施管道，外部用低质木条（9.2cm×120cm×1.5cm）覆盖形成截面面积为40cm×40cm的木质部件。为抵御热带气候的影响，外层选用覆盖型木材。

在被柱子分为四部分的三维空间网格中，木质构件纵横堆叠，垂直向上延伸，人为地创造了"树根"，营造了独立的空间，并为儿童开辟了游乐场所。

"树根"交替堆砌形成了稳定的框架，能够减少室内热量，并通过聚碳酸酯材质的屋顶创造出各种各样的阴影。屋顶中央设有空气层，连接雨水收集池的喷淋系统实现自动化控制，并对空气层进行冷却和净化。

设计方案没有区分结构与覆盖层、天花板与墙面，没有模糊内外之间的界线。该设计充分利用了景观和视野的效果，为顾客带来独特体验，并且通过这样一个蕴含河内街边人行道饮食文化精髓的空间使人们更加地贴近自然。

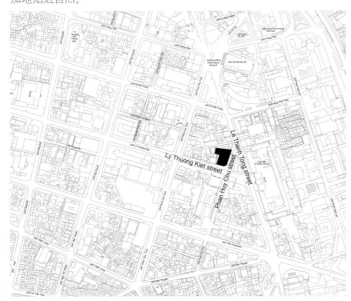

河内街边人行道独有的关键特点
- 天空、太阳、月亮、星星、云彩、风和一排排树木
- 街边人行道——河内的许多日常活动都在这里进行，从早到晚（尤其是烹饪活动）

the keys which make the characteristic of Hanoi sidewalk:
- the sky, the sun, the moon, stars, clouds, the wind and the rows of trees
- sidewalk where many daily activities of Hanoi happened from day to night (especially cuisine)

欢乐餐厅蕴含了河内街边人行道饮食文化的核心价值
Cheering Restaurant contains the core values of the culinary culture of Hanoi sidewalk.

该餐厅是使现代建筑与充满历史和文化价值的文脉产生对话的成功解决方案,同时也有助于重塑河内中心区域的都市形象和建筑风貌——河内寸土寸金的中心区域每平方米地价高达数十亿越南盾。

Cheering Restaurant

Located at the center of Hanoi, Cheering Restaurant was renovated from a long-closed project which still remains steel frame structure and reusable covering materials such as glass, steel, bar steel, sheet-metal roof.

The life on sidewalk, where there are many daily activities of Hanoi people, especially cuisine happens from daytime to nighttime, inspired the designers to create a space that recalls ancient trees – a familiar image of streets of this thousand-year-old city.

The solution proposed welding the entire existing steels to create a new system that can bring technical pipes inside and be covered by low quality wood bars (9.2 cm x 120 cm x 1.5 cm) forming timbers of 40 cm x 40 cm section. Covering wood is selected to resist the tropical climate.

On the three-dimensional network divided by 4, the "timbers" are stacked perpendicularly to each other and gradually spread upwards in order to create "tree roots" for separating spaces and make children playground.

The "tree roots" continue growing alternately to make a stable frame which helps reduce heat and create various shadows from the polycarbonate roof. In the middle of the roof is an air layer which is cooled and cleaned automatically by water spraying system linked to the rainwater collection tank.

The proposal does not divide structure and covering, ceiling and wall. It does not blur the boundary between inside and outside. It thoroughly utilizes views and landscape. It creates special experiences for the users and brings people closer to nature by a space which contains the essence of Hanoi cuisine culture on sidewalk.

The restaurant is a solution to the dialogue between contemporary architecture and the context full of historical and cultural values, while still contributing to reshape the urban and architectural picture in Hanoi central area – where every square meter costs billions VND.

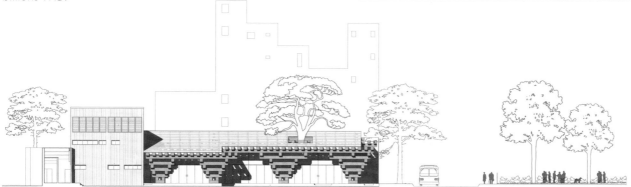

南立面 south elevation

东立面 east elevation 0 5 10m

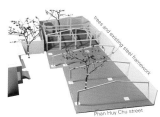

1 原有树木+钢框架
1. existing trees+steel framework

2 第四层木材层+改造后面的建筑
2. 4th wood layer+reforming the building behind

3 第九层木材层
3. 9th wood layer

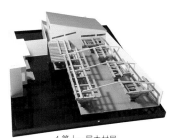

4 第十一层木材层
4. 11th wood layer

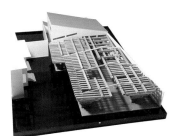

5 第十三层木材层
5. 13th wood layer

6 完成的聚碳酸酯屋面
6. polycarbonate roofing completed

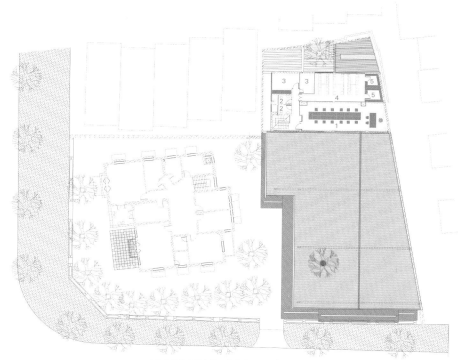

1 办公空间 2 卫生间 3 储藏室 4 员工餐厅 5 更衣室
1. office space 2. restroom 3. storage 4. dining room for staff 5. dressing room
二层 second floor

1 吧台 2 原有树木 3 餐厅 4 游乐区 5 储藏室 6 厨房 7 卫生间 8 原有变电站
9 厨房入口 10 通往二层的通道 11 新种植的树木 12 室外用餐区
1. bar 2. existing trees 3. dining space 4. playing area 5. storage 6. kitchen 7. restroom 8. existing substation
9. the entrance to the kitchen 10. way to second floor 11. new planted trees 12. outdoor dining space
一层 first floor

项目名称：Cheering Restaurant
地点：No. 2 Ly Thuong Kiet street, Hoan Kiem district, Hanoi, Vietnam
建筑师：Doan Thanh Ha, Tran Ngoc Phuong
项目团队：Luu Viet Thang, Chu Kim Thinh, Nguyen Hai Hue, Ho Manh Cuong,
Nguyen Thi Xuyen, Nguyen Gia Phong, Nguyen Van Thinh
承包商：Hexagon.,jsc, Tung Chau.,jsc, Hung Phat.,jsc
用地面积：1700m² / 建筑面积：785m² / 有效楼层面积：1000m²
设计时间：2014.2 / 竣工时间：2014.10
摄影师：©Nguyen Tien Thanh (courtesy of the architect)

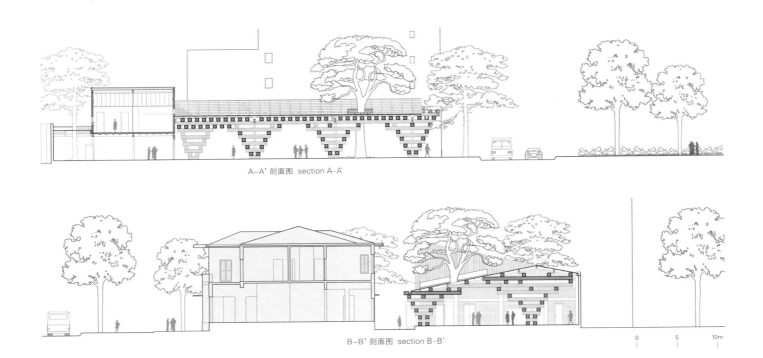

A-A' 剖面图 section A-A'

B-B' 剖面图 section B-B'

紫禁城红墙茶室
CutscapeArchitecture

自然与人工——分为二还是合而为一？ Natural and Artificial - Dichotomy or Duality?

红墙茶室位于紫禁城东南角的一座公共公园内，这座公园曾是明清两代皇帝祭奠祖先的祠堂。祠堂东侧的宫墙作为第二堵围墙环绕着紫禁城，正门是天安门。

在中国的旧社会，围墙曾经长期作为社会阶级划分的象征。而如今，它只是作为空间的划分而存在，墙内是御花园，东面墙外则是百姓居住的胡同住宅。在现代的权力过渡中，一次没有文件记载的毁坏行为推倒了北半部分的围墙。

改造项目在城墙损坏处开展。两间破旧的仓库长久无人使用，一间连接红色宫墙，另一间在第一间北侧3m远处。建筑师设计了一套立竿见影的设计方案，掀掉了第一间仓库的屋顶，露出了胡同一侧的宫墙。一组自由分布的独立钢架结构茶室嵌入露天花园中间，新旧建筑之间形成了多个庭院。北向外观保持相对较好的茶室，外墙用新的覆层材料进行了保温处理，其内部空间被打开以满足现代化的项目需要。通过这种新与旧、实与虚、传统与现代、严肃与荒谬之间的相互影响和作用，对当地环境前所未有的解读方式开始出现了。

项目面临的一大挑战是解决狭窄和密闭问题。首先，在有限的空间之内，必须在满足业主的根本需要的同时，确保不能影响到宫墙内侧的传统文化遗产环境以及宫墙外侧胡同居民的生活条件。设计的基本原则是既要平衡个人隐私，又要共享开放的景观环境；既要创建新的建筑环境，同时又不破坏历史遗迹的完整性。还有一点非常重要，那就是要坚决避免通过模仿传统中国茶室的形象，或在历史遗迹上乱搞奇形怪状的所谓现代风，来伪造传统。

业主寻求的是一处适合在饮茶时间接待商业合作伙伴的、颇具办公氛围的空间。七间风格各异的房间，从3m²到20m²不等，分布于庭院与花园之中，以不同的形状、外观及饰面材料示人。

照明设备及家具的设计与不同的茶室规模保持一致。一系列空间使一度闭塞的环境有了生气，变得活泼起来，也为新建筑环境增色。

建筑师在该项目中尝试使用了多种不同材料，似乎是开了所谓的"极简主义"一个小小的玩笑。这些材料包括石头、竹子和木头等天然材料，钢板、铜和砖块等半合成材料，以及玻璃、聚碳酸酯板和水泥板等一些合成材料。宫墙上深红色的油漆和黄色琉璃瓦，以及胡同住宅斜屋顶上的灰瓦都对建筑材料的选择产生了影响。建筑师充分开发利用了所选材料不寻常的用途，意在延伸它们的潜在用途以赋予这个特殊的地点精巧之感，一如建筑师们以现代化的配置赋予了这座古老的皇宫围墙以全新的视角。

The Forbidden City Red-wall Teahouse

The Red-wall Teahouse is situated inside a public park which used to be the Ancestral Temple – a royal memorial temple for ancestors southeast to the Forbidden City. The palace wall on the east side of the temple wrapped around the Forbidden City as the second ring wall in which Tiananmen functioned as the main gate. The wall had long been a symbol of social class division in old China, now nothing but a spatial threshold separating the impe-

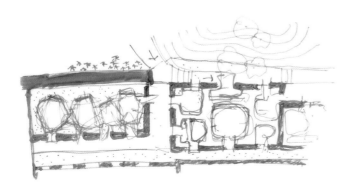

rial garden from the Hutong houses to the east. During modern authority transitions, undocumented damage took down northern half of the wall.

The renovation project took place right at the breakage of the wall. Two shabby warehouses were left on site unattended, one attached to the red imperial wall and the other set around three meters away north to the first one. As an immediate design reaction, the roof of the first warehouse was tipped down to reveal the body of the palace wall on the Hutong side. Then a cluster of freestanding steel-frame tearooms were inserted into the middle of the "unroofed garden" introducing courtyards in between the old and the new structures. The exterior of the north teahouse which had been kept in better shape was insulated with new cover material. Its interior was opened up to match modern program needs. It is, then, through such interplay of old and new, volume and void, tradition and modern, solemnity and absurdness that

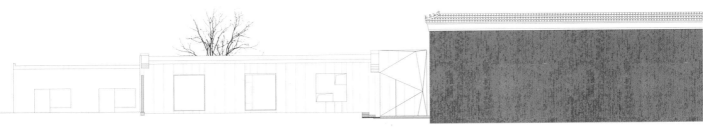

西立面 west elevation

东立面 east elevation

项目名称：The Forbidden City Red-wall Teahouse
地点：The Ancestral Temple, the Forbidden City, Beijing, China
建筑师：Zhang Hong, Zhang Hetian
设计团队：Zhang Cheng, Pan Hongbin, Lian Cheng, Liu Ziyue, An Penghao, Sun Jihua, Jing Jie, Han Xiaowei
供应商：Dasso, Huili, HLLH, etc.
有效楼层面积：280m²
竣工时间：2012—2014
摄影师：
©Zhang Hetian (courtesy of the architect) - p.162, p.171 top-left
©Chen Su (courtesy of the architect) - p.165, p.166, p.167 right, p.170 top
©Wang Yi (courtesy of the architect) - p.164, p.167 left, p.168, p.170 bottom, p.171 top-right, bottom

1 前院　　4 开放坐椅　　7 会议室　　10 储藏室
2 主入口　　5 私人包厢　　8 卫生间　　11 庭院
3 接待处　　6 日光室　　　9 服务室　　12 紫禁城围墙
　　　　　　　　　　　　　　　　　　　13 胡同住宅

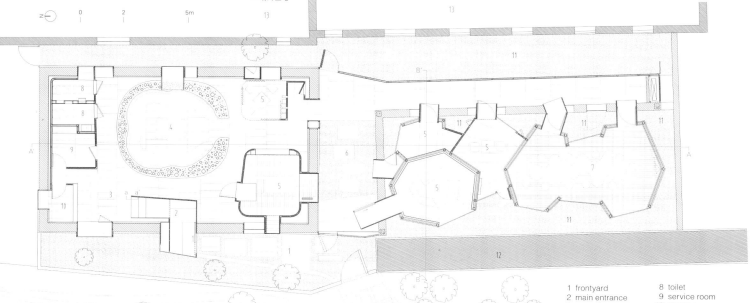

1 frontyard
2 main entrance
3 reception
4 open seating
5 private box
6 sun room
7 meeting room
8 toilet
9 service room
10 storage
11 courtyard
12 forbidden city wall
13 hutong houses

unprecedented readings on local environment start to emerge. Narrowness and closeness were the main challenge of the project. Within the limited space, it is, first of all, essential to meet the program needs of the owner, while at the same time, addressing the heritage environment on the inner side of the palace wall and living conditions of the Hutong dwellers on the other side. To balance individual privacy while sharing open landscape, to establish new built environment while preserving historical legacy were our fundamental tasks. It was also important that we took critical stands against those who falsifies tradition by mimicking an imagery of old Chinese teahouse or those who superimpose arbitrary modern forms over historical heritage.

The owner asked for a quite office space for receiving business partners over tea time. Seven unique rooms, ranging from three to twenty square meters cluster in the middle of courtyards and gardens displaying various shapes, enclosure and finish materials. Lighting and furnishing are designed to correspond to the shifting room scales. The set of spaces starts to enliven the once obstructed environment turning into a contributing part of the new built environment.

In the project we experimented with a variety of materials, which seemed to have made a small joke on so called "minimalism". There are natural materials like stone, bamboo and wood, semi-synthetic materials like steel sheet, copper and brick and synthetic materials like glass, polycarbonate board, cement board, etc.. Dark red paint and yellow encaustic tiles of the palace wall and gray tiles on the pitch roofs of the Hutong houses also exerted influence on the materiality consideration of the project. We exploited uncommon ways of using materials under the determination to extend their potentials for giving delicate sensations to the special place, just like how we rediscover the ancient palace wall in complete modern setting with fresh perspectives. CutscapeArchitecture

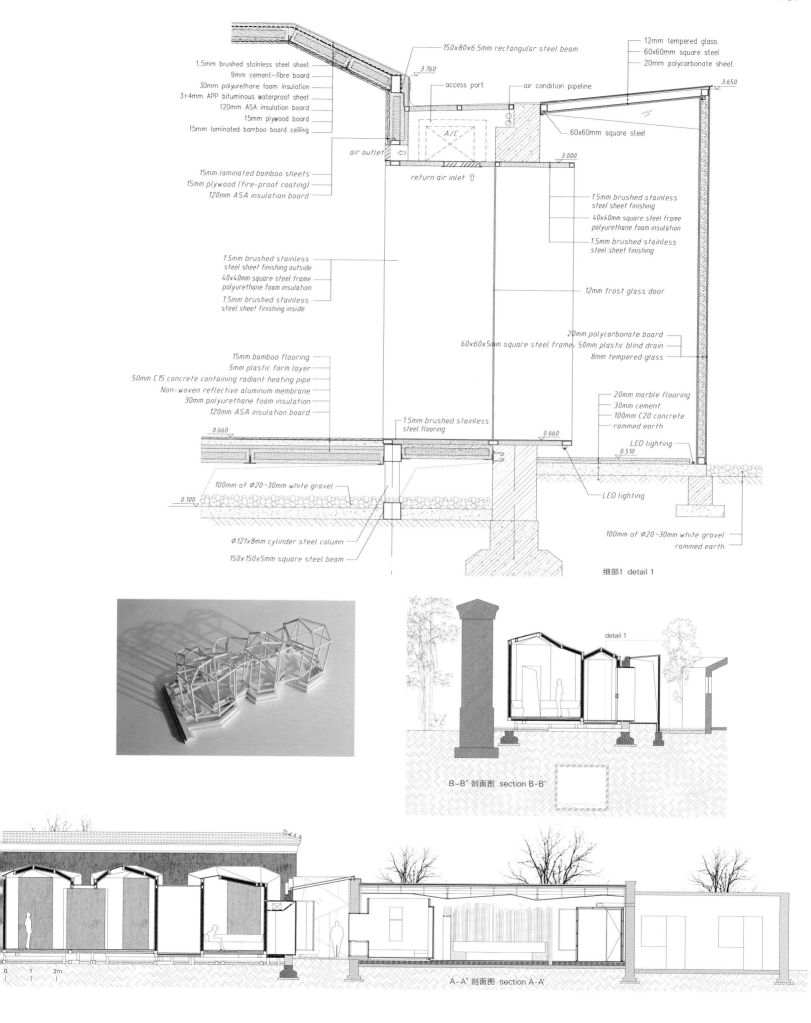

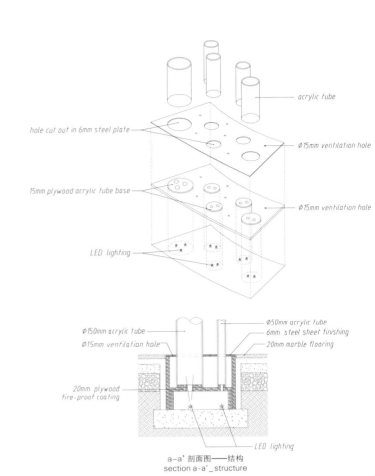

a-a' 剖面图——结构
section a-a'_structure

Steirereck餐厅

PPAG Architects

尽管几年前刚刚经过翻新，但如今Steirereck餐厅的改造设计已然变得尤为必要。业主选择这里的一部分原因在于其位置邻近公园。最终赢得竞赛的设计将单个餐桌的布局作为设计的出发点。特殊的餐桌布局、大型电动窗扇以及有些反光效果并且看起来好似沾有露珠的金属立面，都给新餐厅的顾客一种置身于户外同时又身处家中的感觉。同时，他们体验着最高标准的声学效果和舒适度。建筑师将餐厅立面的材料应用到现有就餐空间的室内，凭借旋转构件创造了所需的大小和比例不一的空间。天花板好似一个悬于就餐空间之上的水平轮廓图，由旋转面板组合而成。餐厅中顾客和餐厅职员会面的中间区域装饰了瓷砖图案，使人联想起厨房的设计，连同装满厨具的橱柜，给顾客一种确实身处美食中心的感觉。地下室的卫生间是餐厅设计中一个很重要的部分，它营造出了一种独特的水晶般的几何结构和氛围，形成一种异位的效果。

餐厅设计产生的效果既新颖又舒适，既能与背景融合，又能体现一种强烈的建筑风格的表达。

Restaurant Steirereck

Despite renovations being completed only a few years ago, a comprehensive reformulation of the restaurant became necessary. The proximity to the park was of particular importance to the clients. The winning competition design takes the individual tables as its starting point. The special table arrangement, the large electric sash windows and the slightly reflective metal facade, which appears to be coated with dew, all give the guests in the new pavilion the sense of being outside and yet also at home. At the same time they experience the highest levels of acoustic and

项目名称：Steirereck im Stadtpark, Vienna
地点：Am Heumarkt 2A, 1030 – Vienna, Austria
建筑师：PPAG Architects Ztgmbh
项目团队：Competition, concept and supervision_Anna Popelka, Georg Poduschka, Lilli Pschill, Ali Seghatoleslami / Project leader_Manfred Karl Botz / Planning team_Roland Basista, Jakub Dvorak, Patrick Hammer, Annika Hillebrand, Philipp Müllner, Lucie Najvarova, Adrian Trifu, Felix Zankel
结构工程师：Werkraum, Wien
机械工程师：Bauklimatik, Wien/Linz
结构框架与木结构：Hazet Bauunternehmung, Wien
衬层与覆层：Individual contractors
用地面积：1950m² / 翻新后楼层面积：2100m² / 有效楼层面积：3000m²
施工时间：2013.9
竣工时间：2014.6
摄影师：©Helmut Pierer (courtesy of the architect)

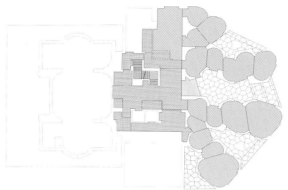

水磨石地砖图案	terazzo tiles/pattern
铸造水磨石	cast terrazzo
门垫	doormat
新瓷砖图案	new tiles/pattern
原有补充瓷砖	existing tiles supplemented

地面覆盖材料
floor cover materials

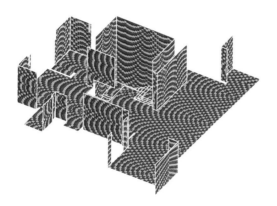

thermal comfort. The material of the pavilion's facade is brought into the interior of the existing dining space, enabling rooms of different sizes and proportions to be created according to needs by means of rotatable elements. The ceiling floats above the dining space like a horizontal contour map, shaped by the possible positions of the rotatable panels. The middle section, where the areas used by guests and personnel meeting, is decorated with a tile pattern reminiscent of a kitchen, which, together with the cabinets filled with kitchen items, gives guests the sense that they are truly involved in the culinary center. The toilets in the basement create a unique crystalline geometry and atmosphere, forming, as a heterotopia, an important part of the restaurant.

The result is something new but also cosy, something that merges into the background but yet is, at the same time, a strong architectural statement.

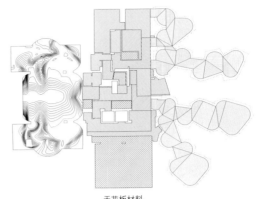

天花板材料
ceiling materials

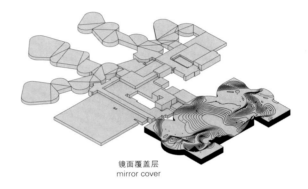

镜面覆盖层
mirror cover

A-A' 剖面图 section A-A'

oak 001	oak 029	oak 087	oak 101
oak 011	oak 071	oak 096	oak 107
oak 014	oak 084	oak 098	oak 108
oak 019	oak 085	oak 100	oak 109

玻璃/镜子 glass/mirror
瓷砖 tiles
铝 aluminium
织物覆盖层 textile covering
木材 wood
涂层 coating
三角形后墙 triangle back wall
裸露混凝土 exposed concrete

墙体材料
wall material

之字形墙体色彩概念
zigzag walls color concept

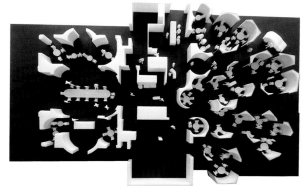

天花板设计图
ceiling development

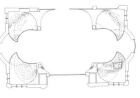

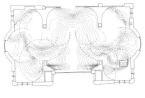
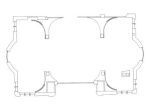

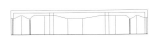
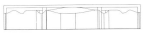

顾客区域不同的设计
guest area variation

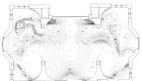
大型 large

大型——15桌 large - 15 tables

大型——10桌+右侧大桌
large - 10 tables + large table right

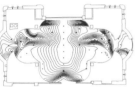
中型 medium

中型——12桌 medium - 12 tables

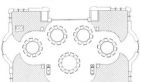
中型——5大桌
medium - 5 large tables

小型 small

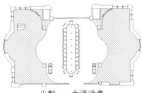
小型——大活动桌
small - event large table

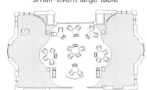
小型——7桌 small - 7 tables

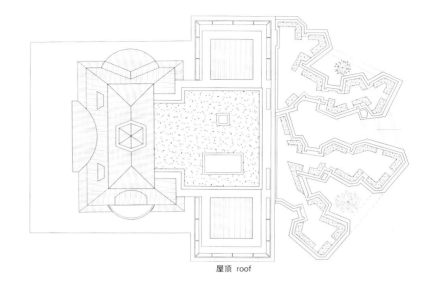
屋顶 roof

一层 ground floor

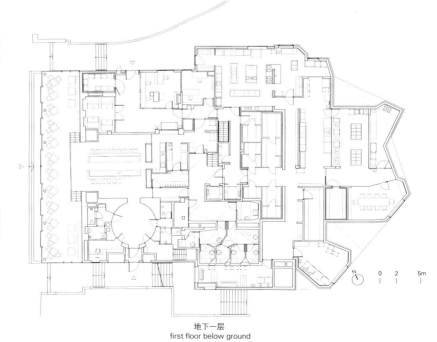
地下一层
first floor below ground

Angolino餐厅
Geneto

自然与人工——分为二还是合而为一？ Natural and Artificial–Dichotomy or Duality?

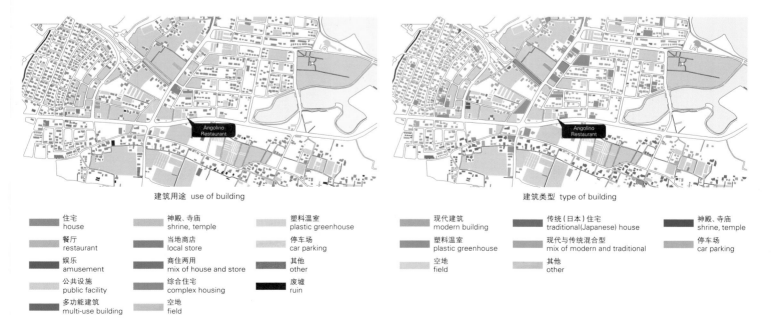

建筑用途 use of building

- 住宅 house
- 餐厅 restaurant
- 娱乐 amusement
- 公共设施 public facility
- 多功能建筑 multi-use building
- 神殿、寺庙 shrine, temple
- 当地商店 local store
- 商住两用 mix of house and store
- 综合住宅 complex housing
- 空地 field
- 塑料温室 plastic greenhouse
- 停车场 car parking
- 其他 other
- 废墟 ruin

建筑类型 type of building

- 现代建筑 modern building
- 塑料温室 plastic greenhouse
- 空地 field
- 传统（日本）住宅 traditional (Japanese) house
- 现代与传统混合型 mix of modern and traditional
- 其他 other
- 神殿、寺庙 shrine, temple
- 停车场 car parking

项目名称：Angolino
地点：Tatebayashi, Gunma, Japan
建筑师：Geneto
结构工程师：Takashi Takamizawa
建筑面积：60.95m²
竣工时间：2013.4
摄影师：©Yasutake Kondo (courtesy of the architect)

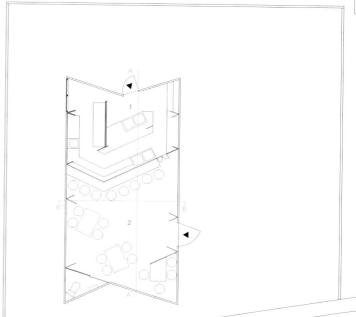

1 厨房	1. kitchen
2 大厅	2. hall
3 独栋住宅（单层建筑）	3. single family house (one-story building)
4 商店（单层建筑）	4. store (one-story building)
5 塑料温室	5. plastic greenhouse

这是一家位于日本群马县的意大利餐厅。相比其他大城市，群马县有着自己典型的城市视野。路边分布着许多连锁商店和大型购物商场。餐厅的业主是本地人，他坚信他的家乡群马县应该具备其自身的特性，并且需要一处可以成为当地人活动落脚点的场所（餐厅）。

建筑师研究了群马县中的建筑形态和用途，尝试设计一家独特的餐厅，并且找机会通过设计一个该地所不具有的建筑形态来改变群马县。餐厅业主认为，该建筑形成了一种面向群马县的设计表达，并且这种表达自建筑的施工阶段就已经开始出现。于是他决定建造一栋自建建筑，并在施工阶段就积极地与由此路过的当地人进行沟通。

在自建和当地不存在此建筑形态的设计前提下，建筑师构思出一个由胶合板组成的简单结构。建筑形态为尖顶形状，以前在该县从没有过这种外形的建筑。它是一个由结构胶合板门式框架（t=24mm）和外墙（t=59mm）组成的硬壳式结构。其外墙由纤维增强塑料防水层覆盖。薄薄的外墙通过光线、声音和气味将室内的气氛带到了户外。

为避免室内外的直接连接，建筑师设计了小三角窗。在室内，呈网状交叉的门式框架将空间和缓地分割开来。厨房和大厅在同一个空间，拉近了大厨和顾客之间的距离。

该建筑物是寻求自建建筑技术水平极限的成果，而建筑立面也在该县产生了一道截然不同的风景线。建筑师希望这家餐厅能够吸引更多的当地人来此聚会，从而使其发展自身的特色，成为一个地标。

Angolino Restaurant

This is an Italian restaurant in Gunma Prefecture, Japan. In contrast to a big city, this town has a typical local city view. There are many chain stores and giant shopping stores in roadside. The client is originally from here and he has a strong thought that his hometown should have identity and it needs a place (restaurant) which can be a foothold for local people.

We researched the building forms and the use in this town. We tried to get identity as a restaurant and make an opportunity to change the town by making a building form not existing here. The client thought that the building has a statement toward the town

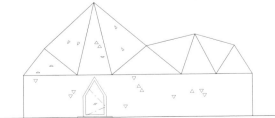
东立面 east elevation

南立面 south elevation

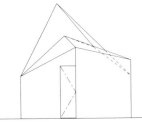
北立面 north elevation

and it arises from the construction stage already. He decided to build as self-build and had a communication actively with the locals passing by during the construction.

On the premise of the self-build and the building form not existing here, we conceived the simple structure by plywood. The building form became the pointed roof shape not existing in this town. It's monocoque construction consisting of the structural plywood portal frame (t=24mm) and the exterior wall (t=59mm). The exterior is covered with FRP waterproof. The thin exterior wall carries the atmosphere of the inside to outside by light, sound and smell.

Avoiding to be connecting the inside and the outside directly, the small triangle windows are built. In inside, the space is separated gently by the portal frame crossing in a net-like form. The kitchen and the hall are the same room to bring a live aspect between the chef and customers.

The structure is the result of looking for the limit of skills by self-build, and the building facade is creating a totally different view in the town. We expect it will develop the identity as a landmark by that the locals get together. Geneto

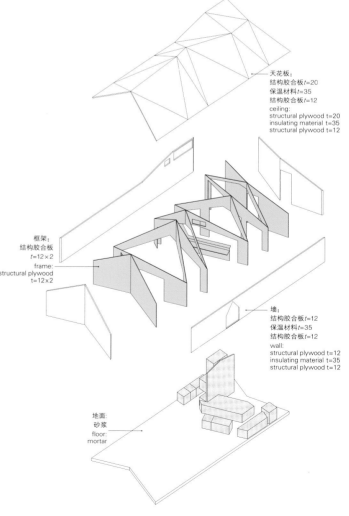

西立面 west elevation

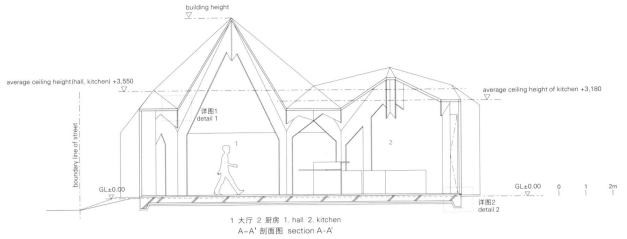

1 大厅 2 厨房 1. hall 2. kitchen
A-A' 剖面图 section A-A'

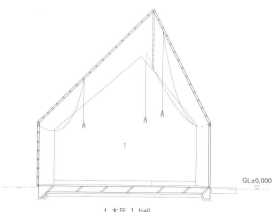

1 大厅 1. hall
B-B'剖面图 section B-B'

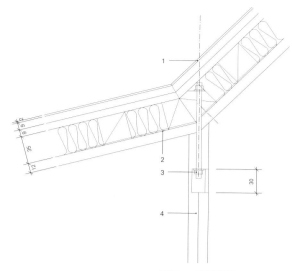

详图1——框架与屋顶
detail 1_frame and roof

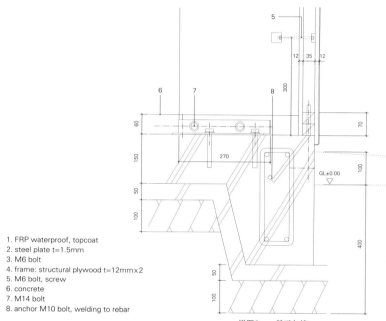

1. FRP waterproof, topcoat
2. steel plate t=1.5mm
3. M6 bolt
4. frame: structural plywood t=12mm×2
5. M6 bolt, screw
6. concrete
7. M14 bolt
8. anchor M10 bolt, welding to rebar

详图2——基础与墙
detail 2_foundation and wall

>>24
Archea Associati
Architects Marco Casamonti[above], Laura Andreini and Giovanni Polazzi were all born in Florence and trained at the School of Architecture of Florence. In 1988, they founded the Studio Archea based in their hometown. In 1999, architect Silvia Fabi also joined the office. In 2001, branches in Rome and Genoa were opened with the participation of Gianna Parisse and Massimiliano Giberti. Besides the main activities of the office which consist of research on design and implementation of architecture at various scales, members of the Archea Associati reconcile intensive occupations of teachers and researchers in several universities in Italy within the area of architectural design.

>>176
CutscapeArchitecture
Zhang Hong and Zhang Hetian are principals of CutscapeArchitecture and GJ Architecture & Engineering. Zhang Hong was born in 1967 at Shanghai, China. Received a B.Arch. from Southeast University and a M.Arch. from Tsinghua University. Zhang Hetian was born in 1976 at Harbin, China and received a M. Arch. from the Metropolitan University of Prague(MUP) and National Superior School of Architecture at Paris-Belleville(ENSA Paris-Belleville), France.

>>60
Collingridge and Smith Architects
It is an award winning international architectural practice, founded by Graham Collingridge and Phil Smith in 2012. Both studied at De Montfort University, Leicester, gaining their BA degree in 1996, received a Diploma in 2000 and further Post Graduate Diploma in Practice Management in 2001. With offices in the United Kingdom and New Zealand, they design and deliver projects that range from refurbishments to new-builds; from domestic scale to urban master plans; from conception to completion. Carefully combines architectural, landscape, interior and furniture design skills to ensure that projects achieve a holistic integrity.

>>168
H&P Architects
Both Doan Thanh Ha and Tran Ngoc Phuong graduated from Hanoi Architectural University in 2002. They set up and have been operating H & P Architects since 2009. In 2014, Blooming Bamboo Home was Highly commended at the WAN House of the Year and AR House Award. They received several awards and certification for innovative architects from Vietnamese Architects Association. They also won the first prize at the International Green Architecture Design Competition and the National Architectural Prize in Vietnam.

>>194
Geneto
Koji Yamanaka graduated from the Kyoto Seika University in 2000. He has taught at his alma mater and the Osaka Seikei University. Yuji Yamanaka studied Living Environment course at Kyoto Prefecture University and received a master degree from the department of Architecture and Building Engineering from Tokyo Institute of Technology in 2006. He has lectured at the Tokyo University of Science, ICS College of Arts and Nippon Institute of technology. They established Geijutsu NinjaTai in 1999, which was renamed Geneto in 2004. Asako Yamashita graduated from the Kyoto prefecture University and completed a master course of Interior Architecture in Konstfack in Sweden. She was a part-time professor in Kyoto Design School. Geneto has received numerous honors such as Good Design Award, JCD Design Award, Design for Asia Award, JID Award Biennial and Architizer A+Awards.

>>48

Petr Hájek Architekti
Petr Hájek studied architecture at the Faculty of Architecture in the Czech Technical University in Prague(CTU) and the Academy of Fine Arts in Prague(AVU), CZ. He founded HŠH Architekti in 1998 and Petr Hájek Architekti (PHA) in 2009. In 2010, he founded Laboratory of Experimental Architecture (LEA) and lectured at the Faculty of architecture in CTU, Prague. He was a leader of the design studio at Academy of Fine Arts and design at Bratislava, Slovakia(VSVU) in 2012. Some of his works were awarded honorable mentions and nominated for Mies van der Rohe Award, aiming on three fundamental fields of interest; designing of buildings, research and education. These topics mutually influenced each other.

João Pedro T. A. Costa
He is an Architect and Master on Contemporary architectonic Culture(TU Lisbon), and a Ph.D. on Urbanism(TU Catalonia, Barcelona). He is a Professor of Urbanism at the School of Architecture, TU Lisbon, where he develops his teaching, research and expertise activity, and Visiting Professor at the Barcelona University(Ph.D. Program), maintaining a general interest in urban and territorial planning issues. His current areas of research are waterfronts, climate change adaptation and urban regeneration.

Aldo Vanini
Practices in the fields of architecture and planning. Mang of his works were published in various qualified international magazines. He is a member of regional and local government boards, involved in architectural and planning researches. One of his most important research interests is the conversion of abandoned mining sites in Sardini.

Paula Melâneo
An architect based in Lisbon. She graduated from the Lisbon Technical University in 1999 and received a master of science in Multimedia-Hypermedia from the cole Supérieure de Beaux-Arts de Paris in 2003. Besides the architecture practice, she also focuses on her professional activity in the editorial field, writing critics and articles specialized in architecture. Since 2001, she has been part of the editorial board of the Portuguese magazine "arqa–Architecture and Art" and the editorial coordinator for the magazine since 2010. She has been a writer for several international magazines such as FRAME and AMC. Participated in the Architecture and Design Biennale EXD'11 as an editor, part of the Experimentadesign team.

>>146

Casos de Casas
It is an architectural innovation program developed by Irene Castrillo Carreira[left] and Mauro Gil-Fournier Esquerra[right] to promote a new agenda for housing in the Mediterranean area. Both are architects from the Madrid School of Architecture, ETSAM. Irene is a free-lance architect focusing on collaboration with renowned and awarded offices such as Carlos Arroyo, Matos-Castillo and Mariano Bayón from Spain. Mauro is a co-founder of Estudio SIC and VIC platform also from Madrid.

>>42

Radionica Arhitekture
It established in 2003 by Goran Rako in Zagreb, Croatia. And today they are compromised by Fani Frković, Dora Krušelj, Klara Nikšić, Jelena Prokop, Ankica Rako, Goran Rako, Ana Ranogajec and Josip Sabolić. In 2004, by winning the first prize in a competition, project Hlapić Kindergarten started its realization. Since then, they have won numerous competitions including Memorial Water-Tower Park and Vučedol Archeological Museum in Vukovar, Cascade Commertial Center in Zagreb and Istanbul Disaster Prevention and Education Center. It was selected as one of five winners of the Collider Activity Center international competition in Bulgaria in 2013.

Wendell Burnette Architects
Wendell Burnette was born in 1962 at Nashville, Tennessee and graduated from the Frank Lloyd Wright School of Architecture, predominantly self-studied and extensively traveled to a lot of countries in Asia, Europe, Africa, South America. He established Wendell Burnette Architects at Phoenix, Arizona in 1996. He is Fellow of the American Institute of Architects and has been teaching at the Design School of Arizona State University since 2000. His works have earned numerous honors, including an Emerging Voices Award from the Architectural League of New York in 1999 and the National AIA Honor Awards in 2007 and 2009, and received the 2009 Academy Award in Architecture from the American Academy of Arts and Letters.

>>134

Patkau Architects
It was founded in 1978 by John Patkau and Patricia Patkau and now operated by four principals including Greg Boothroyd and David Shone. It is an innovative architecture and design research studio based in Vancouver, British Columbia, and Canada. Projects vary in scale from gallery installations to master planning, from modest houses to major urban libraries. As a design leader at Patkau Architects, John has instigated and developed the design of a wide variety of project types for a diverse range of clients nationally and internationally. Whereas Patricia has made important contributions to the field architecture in both practice and education. She is Emerita Professor at the University of British Columbia.

>>72

Henning Larsen Architects
It is an international architecture company in Denmark, with strong Scandinavian roots. It was founded by Henning Larsen in 1959, and is currently managed by CEO Mette Kynne Frandsen and Design Director Louis Becker. It has offices in Copenhagen, Oslo, Munich, Istanbul, Riyadh, the Faroe Islands and a newly established office in Hong Kong. Its goal is to create vibrant, sustainable buildings that reach beyond itself and become of durable value to the user and to the society and culture that they are built into. Its ideas are developed in close collaboration with the client, users and partners in order to achieve long-lasting buildings and reduced life-cycle costs. This value-based approach is the key to there designs of numerous building projects around the world – from complex master plans to successful architectural landmarks. It won the prestigious European Union Prize for Contemporary Architecture-Mies van der Rohe Award 2013.

>>186

PPAG Architects
It was founded by Anna Popelka and Georg Poduschka in 1995. They are great explorers with boundless enthusiasm, always thrilled about anything new. In order to fathom the immanent potential of the three-dimensional, PPAG use logics, science and the game as a matter of course and without any inhibitions. With a great deal of curiosity and ingenuity, they push algorithms, mathematics and aleatoricism to the extreme, applying many elements from their own living environment as tools in the sense of Method Acting, be it dolls, cooking recipes or even their own dwelling as an experimental laboratory – everything can be used for architecture.

>>98

UAO
Architects may become an existence to represent a sudden threat one day to traditional customs and lifestyles. Acknowledging this weightiness of architecture correctly, Mari Ito constructs bold buildings with audacity for greater purpose and simultaneously provides delicate natural beauty of form and ecological and sustainable functionalities. She would like to build "innovation with an affinity" which makes many people feel as "familiar since early times" while accepting – even loving, through her architecture.

>>82

Jorge Mealha
As a portuguese architect, Jorge Mealha was born in 1960 at Maputo, Mozambique, and graduated from the Faculty of Architecture, Technical University of Lisbon in 1987 and successfully concluded his Ph.D. in Architecture in 2013 at Faculty of Arts and Architecture of Lisbon University Lusíada. Since then he has developed projects for a variety of constructions including commercial, housing, industrial premises, schools and sports buildings. He established his own office in 1991 at Lisbon, Portugal and has been currently teaching at the Lisbon University Lusíada since 1991. His works have been published in magazines from Portugal, Spain and Italy. Some projects have been awarded and others been showhed in several exhibitions.

>>160

Atelier KUU
It is a Japanese Architecture, Interior and Lighting design studio which is located at Osaka, Japan. It was founded in 2005 by the architect Nobuo Kumazawa, a former consultant of Takara Space Design. He worked for Takara Space Design for over 30 years after his graduation from the Osaka University of Arts in 1975. Nobuo Kumazawa and Lighting designer, Shiro Miki are currently leading the studio. They won many prestigious Japanese awards such as JCD Desing Award.

>>116

Oller & Pejic Architecture
Monica Oller and Tom Pejic are LA-based native Californian architects, both raised on the West Coast. The regional landscape inspired their education and professional process. Both graduated from the California Polytechnic State University in San Luis Obispo.

C3, Issue 2015.3

All Rights Reserved. Authorized translation from the Korean-English language edition published by C3 Publishing Co., Seoul.

© 2015 大连理工大学出版社
著作权合同登记06-2015年第20号
版权所有·侵权必究

图书在版编目(CIP)数据

景观与建筑：汉英对照 / 韩国C3出版公社编；时真妹等译. —大连：大连理工大学出版社，2015.6
（C3建筑立场系列丛书）
书名原文：C3 Landscaping and Building
ISBN 978-7-5611-9884-1

Ⅰ. ①景… Ⅱ. ①韩… ②时… Ⅲ. ①景观—建筑设计—汉、英 Ⅳ. ①TU-856

中国版本图书馆CIP数据核字(2015)第117142号

出版发行：大连理工大学出版社
　　　　（地址：大连市软件园路80号　邮编：116023）
印　　刷：上海锦良印刷厂
幅面尺寸：225mm×300mm
印　　张：12
出版时间：2015年6月第1版
印刷时间：2015年6月第1次印刷
出 版 人：金英伟
统　　筹：房　磊
责任编辑：许建宁
封面设计：王志峰
责任校对：高　文

书　　号：978-7-5611-9884-1
定　　价：228.00元

发　　行：0411-84708842
传　　真：0411-84701466
E-mail：12282980@qq.com
URL：http://www.dutp.cn